Smart Monitoring and Control in the Future Internet of Things

Smart Monitoring and Control in the Future Internet of Things

Special Issue Editors

Antonio Guerrieri
Franco Cicirelli
Andrea Vinci

MDPI • Basel • Beijing • Wuhan • Barcelona • Belgrade

Special Issue Editors
Antonio Guerrieri
CNR—National Research Council of Italy
Institute for High Performance
Computing and Networking (ICAR)
Italy

Franco Cicirelli
CNR—National Research Council of Italy
Institute for High Performance
Computing and Networking (ICAR)
Italy

Andrea Vinci
CNR—National Research Council of Italy
Institute for High Performance
Computing and Networking (ICAR)
Italy

Editorial Office
MDPI
St. Alban-Anlage 66
4052 Basel, Switzerland

This is a reprint of articles from the Special Issue published online in the open access journal *Sensors* (ISSN 1424-8220) from 2018 to 2019 (available at: https://www.mdpi.com/journal/sensors/special_issues/smart_monitor_IoT).

For citation purposes, cite each article independently as indicated on the article page online and as indicated below:

LastName, A.A.; LastName, B.B.; LastName, C.C. Article Title. *Journal Name* **Year**, *Article Number*, Page Range.

ISBN 978-3-03928-238-8 (Hbk)
ISBN 978-3-03928-239-5 (PDF)

Contents

About the Special Issue Editors

Antonio Guerrieri received his Ph.D. degree in Computer Engineering from the University of Calabria, Italy, in 2012. He is currently Researcher at ICAR-CNR, Italy. He spent six months as Researcher at the Telecom Italia WSN Lab at Berkeley, California, and one year at the Clarity Centre, UCD (University College Dublin), Ireland. He has been involved in several research projects and is a co-founder of SenSysCal S.R.L. His research interests are focused on high-level programming methodologies and frameworks for wireless sensor and actuator networks, building monitoring and control, body sensor networks, design and development of smart environments, smart objects, and Internet of Things.

Franco Cicirelli, Ph.D., has been Researcher at ICAR-CNR (Italy) since December 2015. He was awarded his Ph.D. in System Engineering and Computer Science from the University of Calabria (Italy). He was Research Fellow at the University of Calabria (Italy) from 2006 to 2015. His research work mainly focuses on software engineering tools and methodologies for the modeling, analysis, and implementation of complex time-dependent systems. Other research topics of interest include agent-based systems, distributed simulation, parallel and distributed systems, real-time systems, workflow management systems, Internet of Things, and cyber–physical systems. His research activities involve also Petri Nets, Timed Automata, and the DEVS formalism.

Andrea Vinci, Ph.D., is Researcher at ICAR-CNR, Italy, where he has worked in various positions since 2012. He earned his Ph.D. in System Engineering and Computer Science at the University of Calabria (Italy). His research work mainly focuses on the Internet of Things and cyber–physical systems. In these areas, he has published works on the definitions of platforms and methodologies for the design and implementation of cyber–physical systems, on distributed algorithms for the efficient control of urban drainage networks based on swarm intelligence and peer-to-peer techniques, and on data mining techniques for ambient intelligence.

Preface to "Smart Monitoring and Control in the Future Internet of Things"

The Internet of Things (IoT) and related technologies are promising in terms of realizing pervasive and smart applications which, in turn, have the potential to improve the quality of life of people living in a connected world. According to the IoT vision, all things can cooperate among them and can be managed from anywhere via the Internet, to allow tight integration between physical and cyber worlds, thus improving efficiency, promoting usability, and opening up new application opportunities. Today, IoT technologies are successfully exploited in several domains, providing both social and economic benefits. The realization of the full potential of the next generation of the Internet of Things still needs further research efforts concerning, for instance: the identification of new architectures, methodologies, and infrastructures dealing with distributed and decentralized IoT systems; the integration of the IoT with cognitive and social capabilities; the enhancement of the sensing–analysis–control cycle; the integration of consciousness and awareness in IoT environments; and the design of new algorithms and techniques for managing IoT big data.

This Special Issue gathers contributions and research efforts about advancements in the IoT domain covering important topics such as networking and communication; frameworks and platforms; approaches for modeling, information analysis and discovery; and indoor localization and tracking.

Research outcomes are provided in the fields of smart environments, smart manufacturing, smart health, and smart infrastructures.

Antonio Guerrieri, Franco Cicirelli, Andrea Vinci
Special Issue Editors

Article

Approaching the Communication Constraints of Ethereum-Based Decentralized Applications

Matevž Pustišek *, Anton Umek and Andrej Kos

Faculty of Electrical Engineering, University of Ljubljana, Tržaška 25, SI-1000 Ljubljana, Slovenia; anton.umek@fe.uni-lj.si (A.U.); andrej.kos@fe.uni-lj.si (A.K.)
* Correspondence: matevz.pustisek@fe.uni-lj.si; Tel.: +386-1-4768844

Received: 10 April 2019; Accepted: 6 June 2019; Published: 11 June 2019

Abstract: Those working on Blockchain technologies have described several new innovative directions and novel services in the Internet of things (IoT), including decentralized trust, trusted and verifiable execution of smart contracts, and machine-to-machine communications and automation that reach beyond the mere exchange of data. However, applying blockchain principles in the IoT is a challenge due to the constraints of the end devices. Because of fierce cost pressure, the hardware resources in these devices are usually reduced to the minimum necessary for operation. To achieve the high coverage needed, low bitrate mobile or wireless technologies are frequently applied, so the communication is often constrained, too. These constraints make the implementation of blockchain nodes for IoT as standalone end-devices impractical or even impossible. We therefore investigated possible design approaches to decentralized applications based on the Ethereum blockchain for the IoT. We proposed and evaluated three application architectures differing in communication, computation, storage, and security requirements. In a pilot setup we measured and analyzed the data traffic needed to run the blockchain clients and their applications. We found out that with the appropriate designs and the remote server architecture we can strongly reduce the storage and communication requirements imposed on devices, with predictable security implications. Periodic device traffic is reduced to 2400 B/s (HTTP) and 170 B/s (Websocket) from about 18 kB/s in the standalone-device full client architecture. A notification about a captured blockchain event and the corresponding verification resulted in about 2000 B of data. A transaction sent from the application to the client resulted in an about 500 B (HTTP) and 300 B message (Websocket). The key store location, which affects the serialization of a transaction, only had a small influence on the transaction-related data. Raw transaction messages were 45 B larger than when passing the JSON transaction objects. These findings provide directions for fog/cloud IoT application designers to avoid unrealistic expectations imposed upon their IoT devices and blockchain technologies, and enable them to select the appropriate system design according to the intended use case and system constraints. However, for very low bit-rate communication networks, new communication protocols for device to blockchain-client need to be considered.

Keywords: architecture; blockchain; communication constraints; decentralized application; Ethereum; Internet of things

---

1. Introduction

The Internet o things (IoT) [1] is a well-established concept referring to numerous interconnected things along with corresponding cloud or fog/edge-based applications. It is revolutionizing the Internet and is being deployed in a variety of application domains. The distributed ledgers on the other hand—which are currently mostly implemented with blockchain technologies (BC)—are still emerging [2]. Nevertheless, they are likely to disrupt the field of ICT systems, services, and applications just as strongly as the IoT has in the past. There have already been initial attempts to jointly use the IoT and distributed ledgers [3–10]. These attempts tend to study the feasibility of such application development approaches, provide proofs of concepts (PoC), explore possible use cases, and highlight future business opportunities. Despite the broken illusions about cryptocurrencies in 2018, the technological development in blockchain technologies continues. It started focusing even more on non-monetary applications, including the IoT. This includes seeking performance in BC networks, investigating the role of artificial intelligence for the blockchain, and developing tools, middleware, and service frameworks for real-world business.

The scope of the existing BC systems is divergent in terms of technological features, as well as in their acceptance among the user- and developer communities. In the decade since their first introduction, BC research and development focused on enabling technologies and first BC-based decentralized applications. With the first examples of BC-based IoT solution deployments, certain inefficiencies in current BC designs started appearing. Micropayments for example have become almost unrealistic in the Bitcoin (or Ethereum) network due to high transaction fees and long transaction confirmation times. The scalability and performance needed for the IoT (expected billions of devices) is often limited due to the size of the blockchain and limited transaction rates and excessive latency. The existing BC protocols, e.g., Bitcoin [11] and Ethereum [12], endeavor to face some of these inefficiencies with functional extensions, such as state channels [13,14], sharding [15], and oracles [16]. In parallel, new ledger protocols are being developed, e.g., the Hyperledger Fabric (HLF) [17], NEO [18], IOTA [19], Cardano [20] or Stellar [21] with some of the IoT requirements built-in from scratch. Both developments—the IoT and the BC—are naturally seeking to be combined in common solutions, which thus provide an immense space for application development and use. However, the right approach and the selection of appropriate technologies are far from being straightforward. Beside the differences in technological features, issues like the scalability, transaction costs, deployment in constrained IoT devices must be considered. The same holds for the acceptance of a particular blockchain technology among user and developer communities. The selection can crucially depend on the details of an intended use case, too. Despite a variety of existing and emerging BC technologies, Ethereum is by far the most popular platform for IoT BC applications. Among eleven cases from different application domains presented [22], seven are based on Ethereum and the remaining four are multiplatform (i.e., including the Ethereum).

Interestingly, very little scientific research can be found on actual system and communication requirements to run IoT devices with full blockchain support. For Ethereum even the developers' documentation and installation instructions do not clearly state the necessary allocation of resources to run a full blockchain client successfully. Various user reports [23–25] indicate that at least 2–4 GB RAM and extensive swap sizes are needed to run the client as a full node. For Bitcoin the instructions state just that one needs "a desktop or laptop hardware running recent versions of Windows, Mac OS X, or Linux" and 2 GB RAM [26] for the full client. The storage requirements are determined by the size of the full blockchain, which in May 2019 was about 204 GB in Bitcoin [27], 130 GB in Ethereum with fast sync option applied [28], and 225 GB for a full node [29]. Both are constantly increasing at about 0.1-0.5 GB per day and resulting in approximately 2-5 kB/s of constant communication traffic. In IOTA the requirements are not explicitly defined either, but seem to be comparable to the ones in Ethereum. Block data snapshots are applied in IOTA, which is similar to the pruning concept in blockchain. The storage requirements in IOTA are therefore lower, but can still reach several dozens of GB per node. Despite lacking the precise numbers for the requirements, these figures exceed the capacities of embedded IoT devices.

We therefore face a research and engineering challenge, namely how to bring the blockchain capabilities to the IoT devices that can be constrained in CPU, RAM, storage, communication bandwidth, and energy consumption, and the like. Not all these constraints are necessarily present in every element of an IoT system. The second challenge refers to the security implications of the selected architectural approaches for applications of blockchain technologies for IoT.

In this article we:

(1)   Present the practical constraints from the perspective of end devices in the development of IoT applications based on the Ethereum blockchain as one of the viable and very popular platform for IoT BC applications.
(2)   Elaborate and compare in terms of computation and communication constraints, as well as in terms of security, three architectural approaches for the design of IoT end-device applications based on the Ethereum BC.
(3)   Analyze the results of communication traffic measurements in these architectures to clearly estimate the communication constraints.

Our research provides directions for IoT application designers to enable them to select the appropriate system design and avoiding the placement of unrealistic expectations on IoT devices and BC technologies. Their architectural approach can thus be shaped according to the intended use and the specifics of the planned IoT system.

In Section 2 we briefly present the related work and indicate possible use of decentralized BC applications in IoT. In Section 3 we outline the principles of distributed BC application development for the IoT based on the Ethereum. These principles are elaborated into three architectural approaches for BC enabled IoT devices—Section 4—which differ in communication and computation constraints, as well as in in their security implications. In our pilot installation we measured and analyzed the traffic in various architectures. These results are presented in Section 5 and discussed in Section 6.

2. Related Work

There are not many successful use cases of IoT BC solutions with important and practical business impacts that also reach beyond a proof-of-concept (PoC) and incorporate more than just a limited number of devices. This is not surprising, as the application domain of IoT with blockchain is still in its infancy. Current activities are directed primarily towards the clarification of the role of the BC in the IoT, testing limitations in implementation, and exploring possible business opportunities. Nevertheless, interesting use cases have been presented, primarily in the domains of smart home, smart grid and electric charging, logistics, and IoT device management.

The adoption of BC in IoT is analyzed in [30]. It highlights three significant challenges such as a high resource demand, long latency, and low scalability. It proposes an architecture that combines private and public BC, with simplified block management and cluster headers as gateway entities that provide public BC functionalities for other devices in the system. The same first author in [31] presents the optimization of the BC in the context of smart homes. They analyze security and privacy aspects, along with the overhead introduced by the BC, which remains low and manageable even for resource-constrained devices. Blockchain and IoT integration is further investigated in [32] and in [22].

In [33] they investigate the role of the BC in the mobile charging of electrical vehicles. The concept is presented generally and does not refer to any specific BC technology. The need for the dissemination of chain blocks at all charging stations is not clearly justified and it seems as if a traditional server-based solution could do the job, too. Nevertheless, they point out problems of running full BC clients on constrained nodes. They suggest light clients (Simplified Payment Verification, SPV) and a reduced number of full nodes acting as gateways (called Service Provider, SP).

In [34] they present a prototype of an end-to-end solution, based on an Ethereum BC-controlled IoT electric switch. They implemented the hardware and software for the device, along with the smart contract and the Ethereum compliant web applications for use and control of the system. The device was later upgraded to measure consumption, too. It thus acts as an independent BC node, reporting the measurement status into the chain. In [35] they investigate the support of BCs in various IoT platforms and in [36] they analyze the requirements of IoT devices for the Ethereum BC.

There are other examples of electricity-related use cases of BC technologies for IoT that rely on current public BC networks. In SGs the key challenges that are currently being addressed with the IoT and BC are smart meter reading, selling surplus energy in local microgrids, electric vehicle charging, and demand side management [6,37].

Other application domains present interesting cases of IoT blockchain applications [22], too. Logistics companies are investigating the role of the IoT and BC for product identification and tracking cargo shipments. The idea behind the SmartCargo [7] is that the shipping process should be automated, secure, and transparent throughout the logistic process. Their solution is based on IoT and blockchain and, inter alia, gives access to trustworthy and live cargo tracking. In [4] a container tracking solution is presented that measures light, temperature, and other environmental parameters, and then secures this information in a blockchain. In [5] a similar approach is applied in the pharma supply chain. IoT device management is fundamental to other application domains because it includes access and storage of IoT data in BCs. In [38] this concept is proven in a smart-home scenario to manage home appliances and electricity consumption. A similar idea is elaborated in [39] for the management of vending machines.

Authors in [8] address privacy risks and security concerns in IoT-based healthcare applications. They propose a framework with additional privacy and security properties in a blockchain for IoT, to provide secure management and analysis of healthcare-related big data.

In an envisaged on-demand insurance scenario the study [9] combines blockchain technology with IoT sensors installed in a vehicle. Their proposed system, which is an example of a decentralized blockchain IoT application, enables semi-automatic activation of car insurance coverage.

IoT security, too, can be greatly supported by blockchain technologies. In [40] they elaborate various challenges in effectively implementing security for IoT devices, including resource limitations, device heterogeneity, interoperability of (security) protocols, and scalability and latency of BC networks. Another study [10] elaborates options for access management in IoT. It provides an architecture of a fully distributed access control system based on Ethereum blockchain technology for arbitrating roles and permissions in IoT.

Ethereum as the Ledger Technology for the IoT

In [41] the authors provide a systematic overview of BC technologies and smart contracts for the IoT. They identify several issues that may come up when IoT makers experiment further with BCs, and have their IoT devices participate in a BC network. In this respect they point out the limited transaction throughput (compared to the traditional databases), privacy in the BC, the appropriate selection of BC miners to prevent transaction censoring, the limited legal enforceability of smart contracts, and smart contract validation and security. Rapid developments of blockchain protocols and practical deployments in proofs-of-concepts pointed out that some of the expectations [42] that in the past were placed on blockchain technology for the IoT cannot be taken for granted. Especially current major public BC networks, with their still-present scalability, delay, and cost issues, indicate the need for a clear understanding of new architectural options and required protocol enhancements. An insight into the expectations, actual position and possible remedies is given in Table 1.

The existing BC protocols try to cope with their limitations by making additions that more or less successfully patch the core BC protocols. The state channels, for example, in comparison to the current BC architectures, combine off- and on-chain transactions to contribute to additional scalability, privacy, and the reduction of confirmation delays. In Ethereum this approach is manifested in the Generalized state channels and μRaiden and Raiden networks [13], and in BTC in the Lightning network [14]. The Ethereum smart contracts cannot contact external URLs, which limits their integration with the "world outside of the chain". This shortcoming could be outdone by oracles [16]. These serve as intermediaries, providing data feeds along with an authenticity proof to the blockchain from/to external software (e.g., web sites) or hardware entities. These add-ons have garnered some interest, but are not yet mature (e.g., strong mismatches between announced roadmaps and actual dates of delivery) and with little practical acceptance. This explains why IOTA took a different approach, where the ledger technology (and entire system around it) was designed for the IoT from the very beginning.

Table 1. Expectation put on blockchain technologies in the Internet of things (IoT).

Expectations	Facts	Remedies
Build trust	Yes, trusted transaction ledgers based on decentralized trustless infrastructure.	-
Scalability	Very limited in major public BC networks.	New consensus algorithms; state channels; non-blockchain based distributed ledger technologies; private or permissioned networks, cross-chain solutions
Accelerated transactions	Very poor in major public BC networks, sometimes long transaction confirmation delays and low throughput.	Protocol improvements (state channels, sharding); private or permissioned networks; new BC principles (tangle)
Data monetization, micro payments (cost reduction)	Far from being true in major public BC networks. Transaction costs in the Ethereum are about 0.1 USD and in Bitcoin about 2 USD.	New consensus algorithms; private or permissioned networks
Device autonomy and M2M transactions	Yes, considering the previously mentioned limitations.	-

In this article we elaborate upon the architectures of IoT devices and applications that are based on the Ethereum network. With this selection we in no way wish to single it out as the only appropriate ledger technology in this context. However, the Ethereum network has a proven record in supporting smart contracts, mature implementations of blockchain clients and programming libraries, a large application development community, industry support, and many successful examples of interesting decentralized applications (DApps) already developed. The Ethereum protocols can be implemented in private networks, too, and this can be another option for alleviating, e.g., scalability constraints or transaction costs. For these reasons it is always an option that has to be seriously considered as a viable technological candidate.

The Ethereum protocol, which is being developed by The Ethereum Foundation, is specified in the Yellow paper [43]. New Ethereum transactions are formed in blocks that are validated by mining nodes. The miners use consensus algorithms for validation and they are rewarded for their work. The Ethereum nodes can participate in various public networks, as for example the mainnet or the test network Ropsten.

The key innovation in the Ethereum protocol compared to BTC is the support of smart contracts (SC). These are not some formal requirements or obligations, but can be more adequately explained as autonomous agents, whose behavior is determined by their contract code. This code is executed every time this account receives a message, which is a transaction addressed to it. To develop smart contracts and thus the decentralized applications, a computationally universal (i.e., Turing complete) language is provided. The fundamental smart contract language is the low-level bytecode language and the Ethereum network provides a virtual machine (i.e., Ethereum virtual machine, EVM) which executes such code. Several high(er) level languages are available for application development. The current flagship is Solidity [44]—a JavaScript like language—but other languages have been used in the past. Higher-level code is compiled to bytecode prior to execution in the EVM.

3. Decentralized Applications with Ethereum

Distributed ledgers provide a trusted environment for transactions. In terms of application development for the IoT, two paradigms can be combined—IoT applications with BC support and on-chain logic. Both parts together comprise a decentralized application (DApp), which utilizes the blockchain. Depending on the intended use, both application parts can be combined into one solution, as depicted in Figure 1.

- On-chain application logic refers to smart contracts (i.e., chaincode in Hyperledger Fabric (HLF)), which are programs deployed and executed in the BC network. Executions of smart contracts are validated in BC. BC thus provides a decentralized and trusted virtual machine for smart contract executions. The on-chain logic is not absolutely required for the IoT.

- IoT applications with BC support are web, mobile and embedded applications, which use the BC via client APIs that are made available by the BC clients. These parts of decentralized applications are required for user interfaces and for IoT devices to utilize the BC.

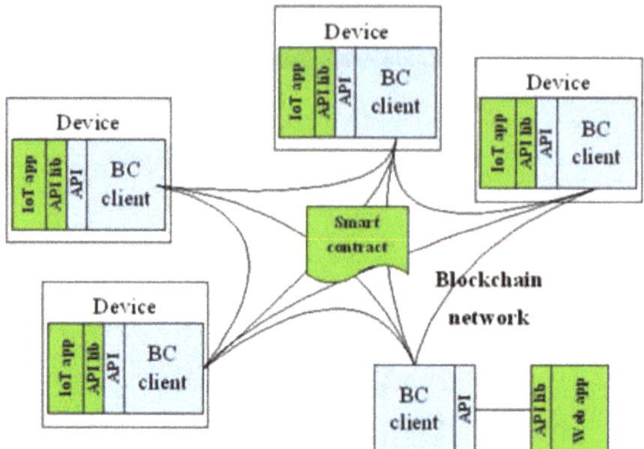

Figure 1. Ethereum's decentralized application architecture for the IoT.

3.1. On-Chain Application Logic

The decentralized environment for trusted transactions, which eliminates the need for trusted central authorities, is the foundation of cryptocurrencies. However, some BC technologies go beyond that and provide smart contracts—the truly revolutionizing feature of the BC, which is not present in traditional web, cloud, and mashup architectures. Smart contracts constitute on-chain business logic that is executed within the blockchain network. Such execution can be verified by any network participant and thus trusted in the same way that any other transaction in a BC network is.

Smart contract code is written in a corresponding programming language (e.g., in Solidity for Ethereum, in Go or Java for HLF, in C#, Java or Python for NEO); it is then compiled to the bitcode suitable for a particular BC, and deployed to the network.

Once deployed in the BC network, a smart contract is addressed by its unique address, similarly to regular BC accounts. A smart contract highlights functions that can be used by other blockchain accounts. These functions represent a kind of an on-chain API for other BC accounts, and are accessible via the blockchain. A smart contract receives transactions addressed to it, with parameters required by a specific SC function embedded in the transaction. The smart contract processes the incoming request according to its programming logic and optionally launches events. The events can be later captured by other clients in the blockchain. The events can thus trigger those actions in IoT applications that rely on the blockchain and have to react upon changes to chain and smart contracts.

3.2. IoT Applications with BC Support

Web, mobile, or embedded applications combine regular application logic (e.g., for user interfaces, sensor data acquisition, local data processing) with BC capabilities. Such capabilities can include a simple transaction exchange in the BC network or communications with an on-chain application part, i.e., a smart contract. The IoT applications use the BC via BC client API libraries and the BC client APIs that are exposed by the BC clients. These functional blocks of the IoT application part are described in more detail in the continuation of this section.

An unmanned embedded IoT system operates without direct user interventions, so a browser is not the appropriate environment for application execution. In that case an application is usually executed in some server side runtime environment (e.g., NodeJS for JS) and the appropriate BC client

API libraries must be imported to the environment for proper operation. This is the foundation of an IoT device with BC support.

There are two key modes of operation for BC-enabled IoT devices to work with and react upon the changes in the BC:

- In the first case a device is identified by a BC address/account. The BC transactions can be sent to and from this address. For the outgoing transactions to be properly signed by the issuers, the location of and secure access to the account key store are needed (see Section 3.3 for details). In this mode the device/application can, e.g., autonomously record its status in the chain.
- In the second case, an IoT device does not have its own BC account. However, even without it, a device can intercept the transactions or the events created by the smart contracts and the BC network. In this way the application can execute certain actions (e.g., toggle on a relay), if a corresponding transaction or event was recorded in the chain (e.g., transaction of some value to a specified BC address). This mode of operation is passive, as IoT devices/application cannot create transactions (operates as a sniffer), but it is much simpler in terms of secure key store management.

While rather distinct in their scopes, both modes of operation have practical value for IoT applications with blockchain support. In smart grids, for example, a passive sniffer could be used for a prepaid energy meter. The meter would intercept its expected status from the blockchain and provide electricity only if enough funds were available. If the consumption exceeds the prepaid quantity, the meter switches off the power. To do this, the meter does not create a transaction and does not need to have its own BC address. An active blockchain node would be required for metering where the device reports its status to the BC network. A meter reading would be reported through a transaction created in the meter and identified by that meters' unique BC address. Any other more complex scenario (e.g., device automation, autonomous negotiations via the BC) requires active nodes, too.

3.3. Building Blocks of an IoT Application for the Ethereum Blockchain

There are five key functional blocks present in an IoT application, with blockchain support to provide the desired functionality and communicate properly with the BC:

- The BC client is responsible for running the BC protocols and thus all communication with the BC network. This includes the management of blocks (keeping the local chain up-to-date) and transactions (e.g., sending outgoing transactions), listening to events, management of peers and the network, monitoring of chain status, managing the accounts or mining blocks. There are several Ethereum BC client implementations available, but geth [45] usually serves as the reference, because it is being developed by Ethereum Foundation developers. A popular alternative is the parity client [46]. There are several synchronization options for the BC client, which affect communication, processing, and storage requirements for the device. Full syncing implies download, verification, and processing of all the chain blocks. In the initial stage fast syncing [28] downloads the transaction receipts along the blocks, and pulls an entire recent state database. Only when the chain reaches a recent enough state, fast sync switches to block processing. This results in much faster synchronization and less download traffic in the initial phase, but potentially opens additional security considerations. With the light syncing option the client only gets the current state. To verify elements, it needs to make inquiries to full nodes. The requirements for light syncing are reduced even further.
- The key store is the location of the private keys associated with a blockchain account. The keys are needed to duly sign the outgoing transactions and thus also access the funds in the account. A lost or stolen key store usually results in severe security breaches.
- The BC client API is a part of the BC client that exposes the clients' capabilities. Through this client API the entire functionality of the BC client can be exploited. The API can be accessed through common programing and communication interfaces, usually the inter-process communication

(IPC), HTTP POST, or Websocket (WS). The IPC can be applied if the application and the BC client run on the same physical device (local communication). The HTTP and WS on the other hand enable also a remote access to the BC client. The data passing through one of these channels is usually structured as JSON.

- BC client API libraries facilitate the application development and use of the BC client API. There are various implementations of these libraries available, for different programming languages and by different developers. In Ethereum such a library is the web3.js [47] (current version 1.0.0) for JavaScript programming. Other implementations may vary in their maturity. These programming libraries are included in the application projects. Apart from interfacing the BC client API, these libraries can provide additional features, as for example a local key store, which keeps and secures access to user accounts and keys, and facilitates the signing of outgoing transactions. This is of utmost importance for the IoT BC application development, as now the application code can manage the accounts easily, securely, and without user interaction.
- The application implements the desired functionality and utilizes the BC through the BC client API libraries.

The application programming code is, in the case of Ethereum, mostly written in JavaScript. The reasons for this are twofold. First, in both cases the JS BC client API libraries are the most advanced and proven, and second, it is suitable for browser-based applications, as well as the IoT device applications.

3.4. Ethereum Transaction Lifecycle

Ethereum transactions transfer value between accounts, pass data to SC function calls, and deploy new smart contracts to the BC network. Once submitted to the BC network, the transactions are organized in blocks by mining nodes, which then execute consensus algorithms upon these blocks. A successfully mined block is added to the chain and the incorporated transactions become part of the irrevocable ledger of past transactions. The Ethereum BC client is responsible for the creation and submission of the transaction to the BC network. An Ethereum BC transaction that is compliant with the Ethereum BC protocol is a set of data that are serialized in the Recursive Length Prefix (RLP) format. In this format there are no parameter names and the input values are in hexadecimal format. Such a message is called a raw transaction.

The BC client API libraries provide functions for signing the transactions and passing them on to the BC network. In the programming language of the application a transaction is presented as a data structure (object), which is then passed on to the corresponding functions. This data structure is unsigned and has no signature filed. In JS programing with the web3.js 1.0.0 library for example, the function sendTransaction() receives such a structure in JSON format, creates the appropriate signature, encodes the results in the RLP format to build the raw transaction, and broadcasts it to the BC network peers. The signTransaction() on the other hand only creates a raw transaction that can be then later passed to the network by, e.g., sendSignedTransaction(). For a transaction to be signed, the sendTransaction() and signTransaction() require access to an unlocked Ethereum account.

3.5. Stand-Alone Blockchain IoT Device Architecture

The stand-alone blockchain device architecture is the initial architecture for the deployment of an end-system (including IoT devices) running applications with blockchain support. The BC client can be configured for full, fast, or light syncing. With full and fast syncing this architecture is only applicable in constraint-less devices and serves as the reference point for elaboration and investigation of practically feasible architectures, which are presented in Section 4. In the stand-alone device architecture, which is depicted in Figure 2, all the functional blocks run on the same physical device. As the BC client (geth) is running there as well, it imposes high demands on the CPU, memory, and storage. If the full BC synchronization is enabled, more than 225 GB [29] of Ethereum chain data must be counted in order to be transferred to and stored at the device. With fast synchronization the size of stored chain data is lower than in full chain, but is still about 130 GB. However, once the client is partially synced, additional growth of the chain data and the related communication traffic are the same in both cases. With light synchronization, communication traffic is further reduced to about 1.1 GB. This setup still requires a reliable and permanent communication channel. The key disadvantage of light syncing is unreliable smart contract event filtering.

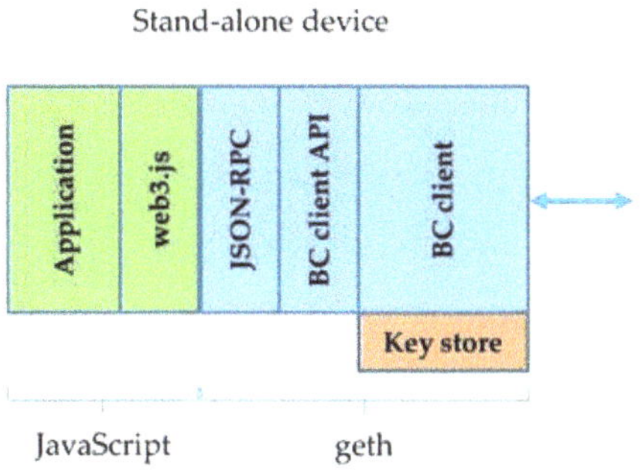

Figure 2. Stand-alone blockchain IoT device architecture [36].

The key store in this case is placed locally and is unlocked by the geth upon the BC client initialization. The key risk in this architecture is the hardware security (stolen keys, if the physical device's privacy is violated).

The constraints from the perspective of a transaction management are summarized in Table 2. Our experience with such a setup showed that it is suitable only for the most powerful (IoT) devices. We tried to run the full client on a Raspberry Pi 3 Model B embedded system with a wired Internet connection. The syncing of the chain proved to be highly unreliable. We experienced unusually long synchronizations (syncing running for several days but still not completing), unexpected interruptions in synchronization, etc. While conducting these tests we had a reference client running on a regular computer (same IP network capacities) and syncing there was unproblematic. It is important to know that an unsynchronized BC prevents the application part from using any BC services. We tried running the geth in light mode, too. The syncing was more successful; however, we experienced severe problems in filtering the events that were launched by our smart contract. Some events were lost due to incomplete data information having been provided, despite the corresponding transactions being duly recorded and chain synced.

Table 2. Possible constraints in an Ethereum transaction management.

Lifecycle Point	Possible Constraints
Creation of a transaction object	No additional constraints
Keeping the key store	Hardware security
Signing a transaction	Computational constraints—might be an issue for very simple devices
Serialization	No constraints
Submitting to the BC network	Communication constraints—if low rate communications are applied
Adding transaction to a chain block	Not relevant for end devices
Syncing the full node, including full and fast mode	Communication and storage constraints— to demanding for an end device
Syncing the light node	Communication constraints—permanent communication channel is required; Application constraints—unreliable smart contract event filtering
Informing about the status of the chain from a remote full client	Communication constraints—if low rate communications are applied

4. Ethereum Application Deployment Options for Constrained Devices

Architectures of IoT applications with blockchain support heavily depend on the capabilities and limitations of the IoT devices where the applications are deployed. The IoT devices demonstrate a wide range of communication (bitrate, persistence of connectivity) and computation (CPU, storage) capabilities, ranging from dumb sensor nodes to fully equipped computers. It is therefore necessary to know these capabilities in advance, to properly select where particular functional blocks (Section 3.3) can run and how they are configured. The challenge is how to organize the required building blocks and implement devices that functionally resemble the stand-alone IoT node—presented in Section 3.5—but with architectures that address possible constraints. All the considerations about the architecture and configurations aimed at providing a reliable execution of the IoT application logic and of the workflow for the Ethereum transactions (creation, signing, submitting, monitoring) and events.

Starting from the stand-alone IoT node architecture, there are two possible architectural choices at disposal:

- the location of the (full) blockchain client, and
- the location of the key store

The first step is to move the blockchain client out of the IoT device and run it on a constraint less network proxy. We call this architecture the remote geth client architecture. Further on we have two possibilities for the key store location. The key store can remain at the IoT device, referred to as a remote geth client with local key store architecture. The second step is to also place the key store to the remote location and keep it with the blockchain client. We refer to this as the remote geth client with remote key store architecture.

Remote geth client architecture to some extent digresses from the fully decentralized peer-to-peer philosophy that is fundamental to the distributed ledgers and BCs. It requires a certain level of trust in the proxy node where the remote client is running. On the other hand, similar approaches are taken, e.g., for most mobile BC clients (that mobile app has to trust the server that provides it the BC functionality). The recent fog computing developments and 4/5G network architectures also indicate that network edge nodes could serve as application gateways, providing functionality to the end nodes. In deployment of decentralized applications we can rely on providers like Infura [48], too. Instead of running Ethereum nodes on our own, a remote node with the API and a reliable BC network connectivity can be provided as a service. In a similar way some other blockchain APIs are made available [49]. As part of the BigQuery Public Datasets program, in 2018 Google Cloud released datasets consisting of the blockchain transaction history for Bitcoin and Ethereum and introduced additional cryptocurrencies later [50]. As the objective of BigQuery is different, this data is meant for BC network and smart contract analytics and not for the deployment of decentralized applications.

However, in addition to the favorable impact on computation, storage, and communication requirements, architectural variations may affect the security of the overall decentralized application. The security impact caused by the architectural variations cannot be completely avoided, but understanding the possible security implications makes the security risks predictable and the requirements acceptable. In particular, the hardware security risk can increase and the level of decentralization can decrease. The possible security implications are therefore also considered in selecting the architectural variations. The stand-alone node clearly provides the highest level of decentralization and the related trust, which is the blockchain's key security attribute. Transferring the blockchain functionalities from the end-device to remote nodes requires trust in these nodes. The level of the required trust varies according to the blockchain services that these nodes are providing. The architectural options in the IoT device design for blockchain range from the (practically infeasible) full nodes, to web-hosted exchanges, where the entire end device utilizes none of the blockchain functions directly. There is less risk, e.g., if a device creates and signs its own transactions and the remote node only distributes them to the blockchain. If the key store is kept at the remote node and the transactions are signed there, then the level of trust in the node must be high. If remote nodes are applied as a service, no application key stores can be kept at the server for security reasons.

The deployment options are compared in Table 3 and analyzed following subsections.

Table 3. Comparison of deployment options.

Requirement/Feature	Standalone Full Node	Remote Client and Remote Key Store	Remote Client and Local Key Store	Remote Client and Proprietary Protocol
Computation	High	Low	Moderate	Low–Moderate [1]
Communication	High	Moderate	Moderate	Low
Storage	High	Low	Low	Low
Decentralization	Highest	Moderate	High	Moderate–High [1]
Device security risk	High	Low	High	Low or High [1]

[1] A range of options is possible.

The remote geth client architecture can be practically deployed in, e.g., smart grid systems. For example, if the power line communications (PLC) is applied for connectivity, the blockchain proxy could reside at a message-aggregating gateway or data concentrator, which is already located at the secondary substation and terminates the PLC connections. A secondary substation usually has its own local area network and broadband connectivity towards higher levels of the grid architecture, so running a full client there is possible.

4.1. Remote Geth Client with Remote Key Store

With remote geth clients and remote key stores we run geth on a separate, constraint-free server. The JavaScript application of course remains at the local IoT device. The most resource intensive part is thus moved from the IoT device. A remote server exposes the geth functionality over JSON-RPC API, with HTTP or WS as the transport channel. In this setup the key store remains at the server. The key store is applied at the geth client initialization, just as in the case of a stand-alone node. The remote geth client with remote key store architecture is depicted in Figure 3.

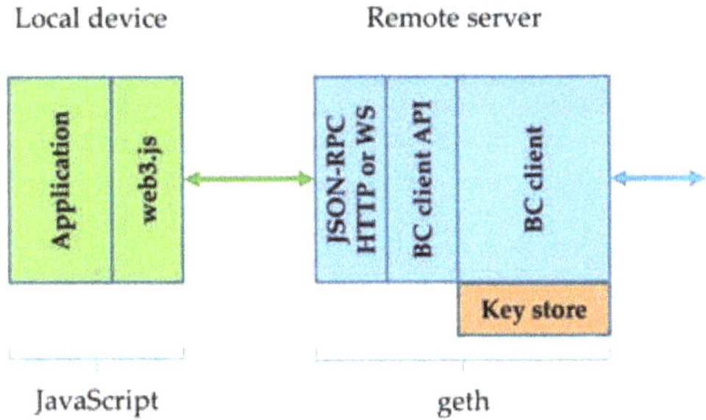

Figure 3. Remote geth client with remote key store [36].

This architecture actually proved to have a practical value. A local device was successful in running the application part, while a remote server seamlessly ran the geth. This approach most efficiently reduces computation and storage requirements set to the local device. It also importantly reduces the communication requirements, because only the incoming notifications about subscribed events and outgoing transactions have to be transferred. With the remote geth client architecture the blockchain synchronization data does not access the local device. As presented later in Section 5, a typical transaction submitted by the application to the geth in the form of JSON-RPC over HTTP was comprised of one HTTP POST request and a corresponding response. In this request the JSON transaction object is passed to the BC client API function call. The size of the request messages was about 800B. The response message was smaller at 280 B. When WS was used instead of HTTP, the messages were roughly 200 B smaller. This does not seem like much, but it can still exceed the communication limits of low bit-rate devices. This is especially true if not merely a limited number of transactions is passed over HTTP/WS, but also some event filtering from web3.js is applied that utilizes the polling principle, generating a constant network load.

However, this architecture has potential security risks we need to understand. If geth is run with the key store unlocked, then anyone accessing the geth with JSON-RPC over HTTP or WS can create transactions signed with this key. There is no access control to HTTP or WS built into the geth, so we need to plan the IP network layer's security very carefully in this case. These risks are relevant only where the IoT device acts as an active transaction creator. If it runs in passive mode (sniffing the chain for transactions and events), then no key store is required, so there are no risks in this respect.

Smart grids are viewed as a possible application of this architecture, predominantly for devices running in passive mode, as for example load control via blockchain or device management.

4.2. Remote Geth Client with Local Key Store

The remote geth client with local key store architecture—which is depicted in Figure 4—and the one with remote key store differ in how the key store is positioned. As this is no longer placed on the server, the security risk of sharing the same geth client among several devices diminishes. In fact, in terms of security, only the client's availability remains a relevant issue. The blockchain data kept by the client is readily available in the BC network to any other participant and is not meant to be private. In this case, however, the application has to (create and) submit raw transactions, including the signature and apply proper serialization. Key store management and signing of transactions is already supported in the web3.js.

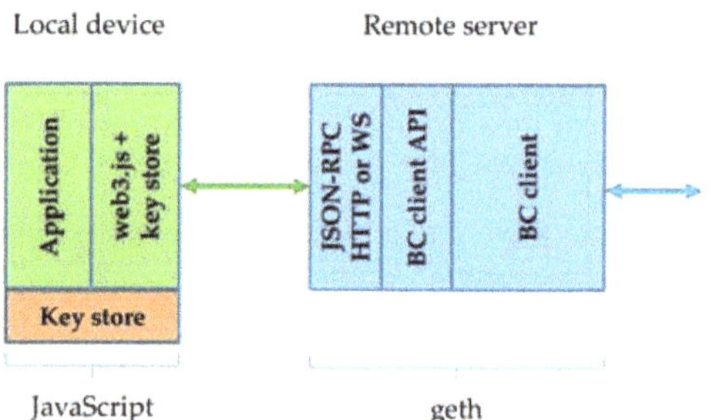

Figure 4. Remote geth client with local key store.

Due to the remote location of the client this approach also efficiently reduces computation and storage requirements set to the local device. The slightly increased computational requirements are related to the signing of transactions, now done locally. It also importantly reduces the communication requirements, just like the architecture with the remote key store. The remaining traffic volumes are comparable to the ones in the remote key store approach, as discussed in Section 4.1. Interestingly, passing raw transactions instead of JSON transaction objects over the HTTP/WS did not result in smaller message sizes, which were expected due to the more efficient RLP encoding. The raw transaction namely includes the signature and transaction hash (not present in JSON transaction object), resulting in messages that in this particular case were about 40B larger.

In terms of security this architecture is much closer to the initial idea of a fully decentralized blockchain network. Every local device is identified by its own Ethereum account and keeps its key store. The need to trust the network proxy is largely reduced and is concentrated to the availability of proxy services. Even a selection of arbitrary geth proxies is possible. However, additional hardware security risks emerge, due to local positioning of the key store and possible device tampering.

Nonetheless, the hardware security requirements are at least to some extent readily addressed in the current (non blockchain) IoT devices, as for example smart meter solutions. Smart meter manufacturers should provide security levels in their devices that are comparable to the ones in online payment systems. This includes secure storage of cryptographic keys and certificates. For example, the study [51] addresses the need for cost-effective tamper-resistant smart energy devices, and [52] the security standards supporting smart grid reliable operation, including the role of trusted computing platforms for smart grid. Nevertheless, it is true that actual secure implementation is still a vendor-specific issue, affected by cost and resource constraints, and performance considerations [53].

4.3. Proprietary Local-Device to Remote-Server Communication

The remote client lets us successfully address the computational and storage constraints. Even though the communication load is drastically reduced compared to the full node architecture, it still remains too big for low-bit-rate networks. The communication between the local device and the geth client over the HTTP/WS was not optimized for minimum communication loads. The redundancy is for example in hexadecimal encodings of raw transactions in API calls, application layer overhead and headers, and event notification management.

As the last option we therefore propose a proprietary communication protocol between the IoT local device end the remote (geth) server, as seen in Figure 5, discarding the existing JSON or RLP data formats. We see two benefits in this; first, the communication bandwidth requirements can be reduced to the minimum, and be thus able to communicate over low bit-rate channels, too. Second, we can apply advanced server access controls to minimize security risks described in the remote geth client with remote key store architecture. According to our knowledge no such solution has been provided for the Ethereum network.

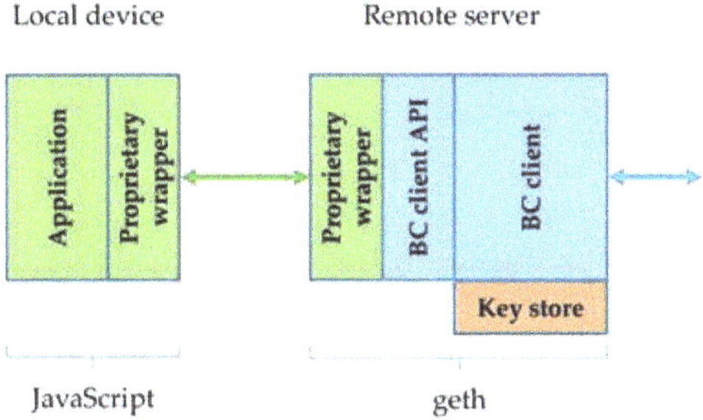

Figure 5. Proprietary wrapper for communication between the device and the remote server.

The approach could be implemented as a proprietary communication wrapper, with corresponding instances placed at the proxy node and at the local device. At the time of writing the web3.js library only facilitates the remote key store approach. Local creation and signing of the transactions with web3.js is only possible if the library has a Websocket connection (which is discarded in this architecture) to a geth client. However, it has been already announced that, in the future, no connection to the geth will be required, so the local key store with proprietary protocols will become viable, too.

This approach can reduce the size of exchanged messages to only several bytes and at the same time does not increase the computation and storage requirements compared to the first two remote geth client architectures. The frequency of the messages can be reduced too. The wrapper at the server does not only have to be transparent, but it can also implement some of the event notification and management logic that is otherwise executed at the local device. Even verification, for example, can be done at the proxy, so only notifications of previously verified events are passed to the local device. Without a proprietary wrapper the device would have to verify the event by issuing additional messages to the proxy.

5. Communication Traffic Measurements and Results

To be able to understand the full scope of possibilities and constraints of Ethereum-based decentralized applications, we had a prototype developed for an Ethereum BC-controlled IoT electric switch and smart meter, dubbed *Swether*, along with smart contracts and Ethereum-compliant web applications [34]. The application deployment options presented in this section and the communication traffic analysis were verified with this pilot DApp setup. We were running the system in the Ropsten public Ethereum test network. The traffic captures and analysis were made with Wireshark. We measured and analyzed the network traffic to the client and the traffic passed between the application and the geth client to get practical insight into possible communication constraints.

5.1. Experimentation Setup and Scenarios

The *Swether* [34] is a prototype smart grid device that can act as an electric switch controlled through the Ethereum network and an energy meter that reports its metering to Ethereum. This requires both key modes of operation for BC-enabled IoT devices. If a device is identified by a BC address/account, the BC transactions can be sent to and from this address. In this case we were able to analyze the transaction-related traffic. If a device operates in passive mode, it only intercepts the event notifications created by the smart contracts or the BC network. In this case we were able to focus on the event notification-related traffic. In both modes there we expected a substantial share of periodic communication traffic needed for blockchain participation, not directly related to application-specific transactions or events. For experimentation purposes we configured a *Swether* in the various deployment options presented in Section 4.

In a typical user scenario a user would book a plug in a *Swether* device for a desired time period or energy quantity, and confirm the booking with a transaction to the Ethereum network sent from the BC-enabled web browser (see Section 3.2 for details). The smart contract would validate the request and launch the events to the blockchain. The events would be intercepted by the *Swether* device. If required, the *Swether* device would report consumption metering via a transaction sent to the smart contact.

The prototype deployment architecture and traffic capturing points are presented in Figure 6. We can distinguish between three traffic categories in this topology, which are labeled in the figure and explained in the continuation.

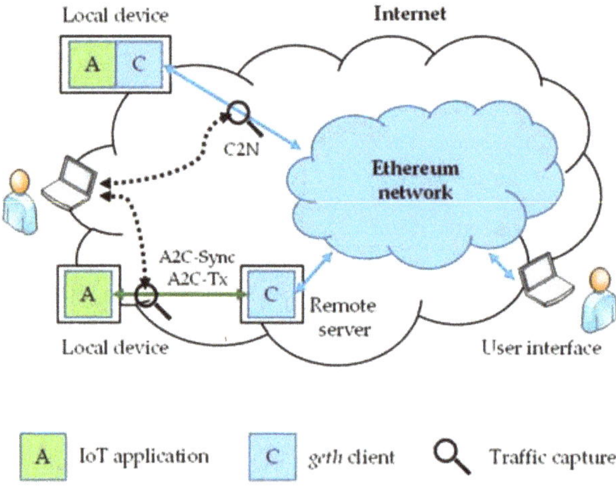

Figure 6. Pilot setup for traffic measurements [nist.sp.1108R3].

For stand-alone end-devices, both the IoT application and the geth client run locally on the IoT device, so the captured traffic includes the entirety of the device-to-Ethereum communication. The application and the client in this case communicate internally, over the IPC. The traffic between the geth client and the Ethereum network (labeled as C2N) is predominantly caused by the periodic blockchain synchronization, but also includes peer-discovery communications, transaction broadcasting, and the like. As application developers, we have little impact on this traffic since it is operated by the geth. An alternative is developing proprietary BC clients, but with that we lose the benefits of any public Ethereum networks.

With remote servers the geth client is moved from the local device and the traffic refers to the JSON-RPC via HTTP or WS (or a proprietary protocol, if implemented) between the device and the remote server. The second traffic category is the periodic traffic between the application and the remote geth client (labeled as A2C-Sync). It results from the client updating the application about the chain status. This is not the chain data as in C2N-Sync, but merely the notifications about new chain block arrivals, so that the application keeps in contact with the client. The third category is traffic related (labeled as A2C-Tx) to the exchange of transactions and event notification between the application and the geth. This traffic is not periodic and is entirely application-specific. The amounts of A2C-Sync and A2C-Tx do not depend on the geth client-syncing mode (full, fast, and light).

5.2. Measurement Results

Table 4 summarizes the results of periodic communication traffic measurement (C2N and A2C-Sync) between the device and the remote geth client.

Table 4. Periodic communication traffic between the device and the geth client.

Architecture	Label	Periodic Traffic
Stand-alone device—full/fast	C2N	18 kB/s
Stand-alone device—light	C2N	1-2 kB/s
Remote geth client with HTTP	A2C-Sync with HTTP	2.4 kB/s
Remote geth client with Websocket	A2C-Sync with Websocket	0.17 kB/s

Table 5 summarizes sizes of transactions to a smart contract address passed from the device to the client.

Table 5. Sizes of transactions to a smart contract address passed from the device to the client.

Communication ->	HTTP		Websocket	
Tx Serialization ->	Raw Tx [1]	Tx Object [2]	Raw Tx [1]	Tx Object [2]
SC data [B]	278	268	278	268
Tx –no SC data [B]	208[3]	178	208[3]	178
Tx -with SC data [B]	486	446	486	446
Request payload [B]	556	511	556	511
Request frame [B]	805	760	618	573
Response payload [B]	103	103	103	103
Response frame [B]	280	280	159	159
web3.js function	sendRawTransaction()	sendTransaction()	sendRawTransaction()	sendTransaction()

[1] Signed transaction passed from the device to the client in RLP serialized form. [2] Transaction object passed from the device to the client in JSON format. [3] Raw transaction encoded as a hexadecimal string.

6. Discussion

6.1. C2N: Traffic between the Geth Client and the Ethereum

In full client architecture with full and fast syncing, the traffic to the geth client from the Ethereum network was measured over a period of about 60 minutes. Already before the capture the client was fully synchronized (i.e., the entire blockchain stored at the client). We disabled any application-related events. The traffic therefore resulted only from the regular synchronization of new blocks. The average data rate was about 18 kB/s and about 10 kB/s in download only. This is a traffic volume that is directly determined by the blockchain network, protocol, and client, and we do not have much influence on it. Even without any additional application-related data, this traffic to the client is needed to keep the local copy of the chain in synchronization.

The light syncing option proved to be inappropriate for our decentralized applications. Despite that, for the reference we measured analyzed the traffic to the geth client with light option. The average data rate was about 1–2 kB/s

6.2. A2C-Sync: Periodic Traffic between the Application and the Geth Client

In a remote client architecture the traffic volume between the node and the full client depends on the applied communication protocol (HTTP or WS), the amount of the data passed to a smart contract in a transaction, and event frequency. Some of the events that result in an exchange of packets are periodic, as for example the indication passed from the client to the node about a new block in the chain. For HTTP this traffic volume was about 2400 B/s and for WS it was 170 B/s. This vast difference results mainly from different modes of operation and not from the applied communication protocol. Less efficient polling is applied for HTTP, and, with WS, the geth client pushes the event indication to the application only when a new block appears.

6.3. A2C-Tx: Transaction and Event Notification-Related Traffic Between the Application and the Geth Client

The appearance of other events is not periodic and results from the DApp's characteristics and use. In our use case each new booking of a charging plug resulted in one captured event and a corresponding confirmation. Besides, the application required one additional event verification after a predetermined number of new blocks in the chain for security reasons. The event capture notification and the corresponding verification resulted in about 2000 B of traffic.

Further, we analyzed the traffic volumes to pass a transaction from the node to the client. A charger would create such a transaction, e.g., to notify the smart contract about the actual amount of energy consumed during a single charging. We considered two cases; with local key store a transaction is created and signed at the node and serialized in the RLP format. Such raw transactions (Raw Tx) are passed by the client to the Ethereum network. For remote key store the node passes a JSON transaction object to the client, which then signs, serializes, and broadcasts it to the network. The sizes of packets for both cases are shown in Table 5. Both the HTTP and WS communication protocols were applied. There is the web3.js function provided in the table to indicate how the transaction was built.

The sizes given in Table 5 refer to the transactions that include some data for the smart contract. The minimum size of a raw transaction (no communication headers included) with no additional data was 104 B. However, when a raw transaction is submitted to the client, it is converted into a hexadecimal string, which is the actual parameter in the client API function call. The string encoding duplicates the size of the raw transaction in bytes to 208 B. Additional data for the SC function call can be included in the transaction. In our case the size of this data was 278 B and 268 B. Beside these fields to build the required JSON, input for the API function is also added. We can estimate that the entire request frame is about 300–500 B if there is no SC data. The SC data size is added up to the frame size with no data. Request frame sizes in HTTP are about 200 B bigger than in WS due to more compact application layer headers. The size of the frame with a JSON transaction object is 45 B smaller than a corresponding raw transaction. While the raw transaction was expected to be serialized more efficiently due to RLP, that was not the case. The increased size of frame with a raw transaction results from the additional RLP to string encoding and from the signature, which is added to the raw transaction (but is not a part of the JSON Tx object).

7. Conclusions

Blockchain technologies are attracting immense attention, both positive and negative, mostly due to current hectic activities in cryptocurrency markets and ICOs. Our research is focused on the technical aspects of blockchain technologies. In our view, which is backed by the initial research, developments, and application examples, there is a vast IoT application opportunity for these technologies, especially in relation to smart grids and smart energies. Our research provides directions for IoT application designers to enable them to select the appropriate system design and avoiding unrealistic expectations imposed to the IoT devices and BC technologies. The architectural approach can be thus shaped according to the intended use and the specifics of the planned IoT system.

We investigated how to match the requirements and constraints of the IoT devices found, e.g., in the smart grid customer domain, as for example smart meters, smart grid gateways, or data concentrators. In standalone device architecture, the application and the blockchain client run on the same device. This imposes computational, storage, and communication requirements that smart meters or gateways cannot meet, and makes the implementation of blockchain nodes for smart energy as standalone end-devices impractical or even impossible. We measured and analyzed the communication traffic of a standalone blockchain node and compared it to the traffic where the blockchain client was moved to a remote server. We found out that it is possible to distinguish between two traffic categories. Periodic communication traffic is needed for blockchain participation, i.e., syncing the blockchain or receiving event notifications about synchronization. Traffic that results from event notifications and transaction exchange is not periodic and depends on application operation. For application operation we therefore analyzed the sizes of particular transactions instead of the average load. The periodic traffic of a standalone Ethereum node is about 18 kB/s.

The most resource-demanding part in the remote server architecture—i.e., the geth client—is moved from the end-device and placed on a server. This significantly reduces the computation and storage requirements, which can now be met by, e.g., current smart meters. The periodic traffic of a device is strongly reduced from the initial 18 kB/s, too. For HTTP it is about 2400 B/s and with Websocket about 170 B/s. The difference in the two is mostly due to inefficient polling applied by the geth client in the case of HTTP.

A notification about a captured event and the corresponding verification resulted in an exchange of about 2000 B. A transaction sent from the application on the device to a remote client resulted in a message of about 500 B with HTTP and in 300 B with Websocket, due to the more compact application layer headers. If there are additional smart contract input data in the transaction, they are added to the values above. The size of smart contract data is entirely application-dependent. The key store location, which affects the serialization of the transaction, only had a small influence on the transaction-related data. With raw transactions the messages were 45 B larger that when passing the JSON transaction objects. The reason for this is the transaction signature, which is included in the raw transaction. This is a positive finding, since the placement of the key store can be now selected predominantly based on the security requirements and not to meet the communications constraints.

To glean additional insight into our results, we mapped our measurements to estimated NB-IoT and LoRaWAN client conditions. For NB-IoT with a peak rate 250 kb/s and 100 connected end devices, the available bitrate for one end device results in an average rate of 2.5 kb/s (or 312 B/s). This means that the node could be kept synchronized if WS and remote client architecture were applied. The periodic traffic would use 55% of the available bitrate. A remote node with HTTP and a full node would exceed the available bitrate in this case. If all the remaining bitrate (12 MB/day) were dedicated to blockchain transactions, this would additionally allow for transfer of about 40,000 transactions. LoRaWAN is not appropriate for periodic traffic, because of impermanent connectivity and excessive traffic even for WS. It could be suitable to transfer transactions and event notifications. In LoRaWAN connectivity, 100 connected clients and 1% duty cycle result in 70 seconds of air time per device per day. This is about 800 kb/day or 100 kB/day per device. This means that about 333 transactions or 50 event notifications per day could be transferred for the device.

Remote server architecture helps reduce the communication constraints, but the traffic still exceeds the capacities of narrow-band low-bit rate networks. This indicated the direction of future expansion of our research. We are therefore currently specifying and implementing a low bit-rate Ethereum (LBE) protocol for communication from proprietary device to remote client, with the ambition to test it in LoRaWAN and other low bitrate networks. We expect to further reduce periodic traffic volumes, so as to manage the number and limit the size of event notifications, and reduce transaction object sizes. We also plan to set up blockchain nodes for the Ethereum, Bitcoin, IOTA, and NEO networks to continuously monitor the actual system requirements and make these findings available to the research community and thus help to overcome the vaguely defined system requirements for blockchain devices.

Author Contributions: Conceptualization, M.P. and A.K.; Methodology, M.P. and A.U.; Investigation, M.P. and A.U.; Resources, M.P.; Writing—Original Draft Preparation, M.P.; Writing—Review & Editing, M.P., A.U. and A.K.

Funding: The authors acknowledge the financial support from the Slovenian Research Agency (research core funding "Algorithms and Optimization Procedures in Telecommunications", No. P2-0246-1538-04).

Conflicts of Interest: The authors declare no conflict of interest. The founding sponsors had no role in the design of the study; in the collection, analyses, or interpretation of data; in the writing of the manuscript, and in the decision to publish the results.

References

1. Dechamps, A.; Duda, A.; Skarmeta, A.; Lathouwer, B.D.; Agostinho, C.; Cosgrove-Sacks, C.; Doukas, C.; Pastrone, C.; Tragos, E.; Ibanez, F.; et al. *Internet of Things Applications—From Research and Innovation to Market Deployment*; Vermesan, O., Friess, P., Eds.; River Publishers Series in Communications; River Publishers: Delft, The Netherlands, 2014; ISBN 978-87-93102-94-1.
2. The 4 Phases of the Gartner Blockchain Spectrum. Available online: https://www.gartner.com/smarterwithgartner/the-4-phases-of-the-gartner-blockchain-spectrum/ (accessed on 10 April 2019).
3. Share & Charge–Charging Station Network–Become Part of the Community! Available online: https://shareandcharge.com/en/ (accessed on 9 April 2019).
4. Ford, N. IoT Application Using Watson IoT & IBM Blockchain. Available online: https://www.mendix.com/blog/built-iot-application-10-days-using-watson-iot-ibm-blockchain/ (accessed on 9 April 2019).
5. Bocek, T.; Rodrigues, B.B.; Strasser, T.; Stiller, B. Blockchains everywhere–a use-case of blockchains in the pharma supply-chain. In Proceedings of the 2017 IFIP/IEEE Symposium on Integrated Network and Service Management (IM), Lisbon, Portugal, 8–12 May 2017; pp. 772–777.
6. Brooklyn Microgrid|BMG 101. Available online: https://www.brooklyn.energy/bmg-101 (accessed on 9 April 2019).
7. SmartCargo. Available online: https://www.smart-cargo.org/ (accessed on 13 May 2019).
8. Dwivedi, A.D.; Srivastava, G.; Dhar, S.; Singh, R. A Decentralized Privacy-Preserving Healthcare Blockchain for IoT. *Sensors* **2019**, *19*, 326. [CrossRef] [PubMed]
9. Lamberti, F.; Gatteschi, V.; Demartini, C.; Pelissier, M.; Gomez, A.; Santamaria, V. Blockchains Can Work for Car Insurance: Using Smart Contracts and Sensors to Provide On-Demand Coverage. *IEEE Consum. Electron. Mag.* **2018**, *7*, 72–81. [CrossRef]
10. Novo, O. Blockchain Meets IoT: An Architecture for Scalable Access Management in IoT. *IEEE Int. Things J.* **2018**, *5*, 1184–1195. [CrossRef]
11. Protocol Documentation–Bitcoin Wiki. Available online: https://en.bitcoin.it/wiki/Protocol_documentation (accessed on 9 April 2019).
12. Viktor Trón; Felix Lange Ethereum Specification. Available online: https://github.com/ethereum/go-ethereum/wiki/Ethereum-Specification (accessed on 9 April 2019).
13. The Raiden Network. Available online: http://raiden.network/ (accessed on 9 April 2019).
14. Lightning Network. Available online: http://lightning.network/ (accessed on 9 April 2019).
15. Vitalik Buterin; Tomoya Ishizaki on Sharding Blockchains. Available online: https://github.com/ethereum/wiki/wiki/Sharding-FAQ (accessed on 9 April 2019).
16. Oraclize Documentation. Available online: http://docs.oraclize.it/#overview (accessed on 9 April 2019).

17. IBM Blockchain Based on Hyperledger Fabric from the Linux Foundation. Available online: https://www.ibm.com/blockchain/hyperledger.html (accessed on 9 April 2019).
18. NEO White Paper. Available online: http://docs.neo.org/en-us/index.html (accessed on 9 April 2019).
19. IOTA for Developers. Available online: https://www.iota.org/get-started/for-developers (accessed on 9 April 2019).
20. Cardano Foundation. Available online: https://cardanofoundation.org/en/ (accessed on 9 April 2019).
21. Stellar Development Guides. Available online: https://www.stellar.org/developers/guides/ (accessed on 9 April 2019).
22. Reyna, A.; Martín, C.; Chen, J.; Soler, E.; Díaz, M. On blockchain and its integration with IoT. Challenges and opportunities. *Future Gener. Comput. Syst.* **2018**, *88*, 173–190. [CrossRef]
23. Ordine, A. Ethereum Client Platforms: Parity Versus Go-Ethereum (19) Ethereum Client Platforms: Parity Versus Go-Ethereum|LinkedIn. Available online: https://www.linkedin.com/pulse/ethereum-client-platforms-parity-versus-go-ethereum-andrei-ordine/ (accessed on 9 April 2019).
24. Ethereum Node Setup on A Virtual Server. Available online: https://github.com/bokkypoobah/BokkysCheatsheet/wiki/Ethereum-Node-Setup-On-A-Virtual-Server (accessed on 9 April 2019).
25. The System Requirements are Seriously Under Estimated. Available online: https://github.com/paritytech/parity/issues/4635 (accessed on 9 April 2019).
26. Running A Full Node–Bitcoin–Minimum Requirements. Available online: https://bitcoin.org/en/full-node#minimum-requirements (accessed on 9 April 2019).
27. Bitcoin, Litecoin, Namecoin, Dogecoin, Peercoin, Ethereum Stats. Available online: https://bitinfocharts.com/ (accessed on 9 April 2019).
28. Fast Synchronization Algorithm. Available online: https://github.com/ethereum/go-ethereum/pull/1889 (accessed on 9 April 2019).
29. Ethereum Sync (Default) Chart. Available online: https://etherscan.io/chartsync/chaindefault (accessed on 14 May 2019).
30. Dorri, A.; Kanhere, S.S.; Jurdak, R. Blockchain in internet of things: Challenges and Solutions. *arXiv* **2016**, arXiv:1608.05187.
31. Dorri, A.; Kanhere, S.S.; Jurdak, R.; Gauravaram, P. Blockchain for IoT security and privacy: The case study of a smart home. In Proceedings of the 2017 IEEE International Conference on Pervasive Computing and Communications Workshops PerCom Workshops, Athens, Greece, 19 August 2017; pp. 618–623.
32. Panarello, A.; Tapas, N.; Merlino, G.; Longo, F.; Puliafito, A. Blockchain and IoT Integration: A Systematic Survey. *Sensors* **2018**, *18*, 2575. [CrossRef] [PubMed]
33. Kim, N.H.; Kang, S.M.; Hong, C.S. Mobile charger billing system using lightweight Blockchain. In Proceedings of the 2017 19th Asia-Pacific Network Operations and Management Symposium APNOMS, Seoul, Korea, 27–29 September 2017; pp. 374–377.
34. Pustišek, M.; Bremond, N.; Kos, A. Electric Switch with Ethereum Blockchain Support. *IPSI TIR* **2018**, *14*, 21–28.
35. Pustišek, M.; Štefanič, L.J.; Kos, A. Blockchain Support in IoT Platforms. *IPSI TIR* **2018**, *14*, 13–20.
36. Pustišek, M.; Kos, A. Approaches to Front-End IoT Application Development for the Ethereum Blockchain. *Procedia Comput. Sci.* **2018**, *129*, 410–419. [CrossRef]
37. Wang, J.; Wang, Q.; Zhou, N.; Chi, Y. A Novel Electricity Transaction Mode of Microgrids Based on Blockchain and Continuous Double Auction. *Energies* **2017**, *10*, 1971. [CrossRef]
38. Huh, S.; Cho, S.; Kim, S. Managing IoT devices using blockchain platform. In Proceedings of the 2017 19th International Conference on Advanced Communication Technology ICACT, PyeongChang, Korea, 19–22 February 2017; pp. 464–467.
39. IBM Watson Internet of Things Blockchain and IoT: Vending Machine with eSIM Demo. Available online: https://www.youtube.com/watch?v=T9kYuBcOnjI (accessed on 9 April 2019).
40. Khan, M.A.; Salah, K. IoT security: Review, blockchain solutions, and open challenges. *Future Gener. Comput. Syst.* **2018**, *82*, 395–411. [CrossRef]
41. Christidis, K.; Devetsikiotis, M. Blockchains and Smart Contracts for the Internet of Things. *IEEE Access* **2016**, *4*, 2292–2303. [CrossRef]
42. Pureswaran, V.; Brody, P. *Device Democracy—Saving the Future of the Internet of Things*; IBM: New York, NY, USA, 2015.

43. Gavin Wood The "Yellow Paper": Ethereum's Formal Specification. Available online: https://ethereum.github.io/yellowpaper/paper.pdf (accessed on 9 April 2019).
44. Solidity–Solidity 0.5.x Documentation. Available online: http://solidity.readthedocs.io/en/latest/index.html (accessed on 9 April 2019).
45. Viktor Trón; Felix Lange Geth. Available online: https://github.com/ethereum/go-ethereum/wiki/geth (accessed on 9 April 2019).
46. Parity Technologies. Available online: https://parity.io/ (accessed on 9 April 2019).
47. Web3.js–Ethereum JavaScript API. Available online: https://github.com/ethereum/web3.js (accessed on 9 April 2019).
48. Infura–Scalable Blockchain Infrastructure. Available online: https://infura.io (accessed on 13 May 2019).
49. BlockCypher Developer Portal—Bitcoin, Ethereum, and Blockchain APIs. Available online: https://www.blockcypher.com/dev/ (accessed on 13 May 2019).
50. Introducing Six New Cryptocurrencies in BigQuery Public Datasets–and How to Analyze Them. Available online: https://cloud.google.com/blog/products/data-analytics/introducing-six-new-cryptocurrencies-in-bigquery-public-datasets-and-how-to-analyze-them/ (accessed on 15 May 2019).
51. The Smart Grid Interoperability Panel–Smart Grid Cybersecurity Committee. *Guidelines for Smart Grid Cybersecurity*; National Institute of Standards and Technology: Gaithersburg, MD, USA, 2014.
52. *Smart Grid Information Security*; *SG-CG/M490/H*; CEN-CENELEC: Brussels, Belgium, 2014; p. 95.
53. *Essential Regulatory Requirements and Recommendations for Data Handling, Data Safety, and Consumer Protection*; European Commission: Brussels, Belgium, 2011; p. 123.

Article

Computation of Traffic Time Series for Large Populations of IoT Devices

Mikel Izal [1,2,*], Daniel Morató [1,2], Eduardo Magaña [1,2] and Santiago García-Jiménez [1]

[1] Electrical, Electronic and Communications Engineering Department, Universidad Pública de Navarra, 31006 Pamplona, Spain; daniel.morato@unavarra.es (D.M.); eduardo.magana@unavarra.es (E.M.); santiago.garcia@unavarra.es (S.G.-J.)

[2] Smart Cities Institute, Universidad Pública de Navarra, 31006 Pamplona, Spain

* Correspondence: mikel.izal@unavarra.es; Tel.: +34-948-169-838

Received: 24 October 2018; Accepted: 21 December 2018; Published: 26 December 2018

Abstract: The Internet of Things (IoT) contains sets of hundreds of thousands of network-enabled devices communicating with central controlling nodes or information collectors. The correct behaviour of these devices can be monitored by inspecting the traffic that they create. This passive monitoring methodology allows the detection of device failures or security breaches. However, the creation of hundreds of thousands of traffic time series in real time is not achievable without highly optimised algorithms. We herein compare three algorithms for time-series extraction from traffic captured in real time. We demonstrate how a single-core central processing unit (CPU) can extract more than three bidirectional traffic time series for each one of more than 20,000 IoT devices in real time using the algorithm DStries with recursive search. This proposal also enables the fast reconfiguration of the analysis computer when new IoT devices are added to the network.

Keywords: IoT; network traffic; monitoring; DDoS; packet classification

---

1. Introduction

The Internet Protocol (IP) provides connectivity to millions of smart and autonomous devices. They range from small health monitors, ambient sensors, and location notification devices to seismic sensors, traffic cameras, and generic computers. The smart devices represent a new wave in network-connected elements that is expected to surpass the number of computer hosts. Different predictions estimate between 20 to 200 billion of these devices by 2020 [1,2].

The service architecture for each type of device differs, but a centralised information-collecting element is typical. Deployed devices typically communicate with a central collector using virtual private networks offered by mobile communication companies or local Internet service providers (ISPs) [1]. An example scenario would be energy consumption-metering devices deployed in households; they communicate with the central office using power lines. Other examples include sensors deployed in remote wind power production towers that communicate using cellular networks, temperature and pressure sensors for weather forecasting, and location devices for fleet tracking logistics [3]. The Internet of Things (IoT) ecosystem offers a plethora of examples of large populations of small sensing devices that collect information and send them to centralised hosts.

The process of monitoring the availability of these 'things' is a difficult task, owing to the large number of devices. It can be achieved by active or passive monitoring. Any type of check that actively communicates with the devices would inject a large amount of traffic into the access networks, whereas passive monitoring techniques do not add any load to the network or the devices. Passive monitoring can be based on a time-series analysis of network traffic from each device. These traffic time series can be used to verify device liveliness by detecting periods of network traffic silence (Figure 1). For cellular operators, traffic time series are fundamental for evaluating traffic patterns from different types of

IoT devices. It is noteworthy that these devices typically employ a cellular network [4], competing for resources with smartphone users [5]; therefore, network dimensioning requires traffic profiles for different cellular user types.

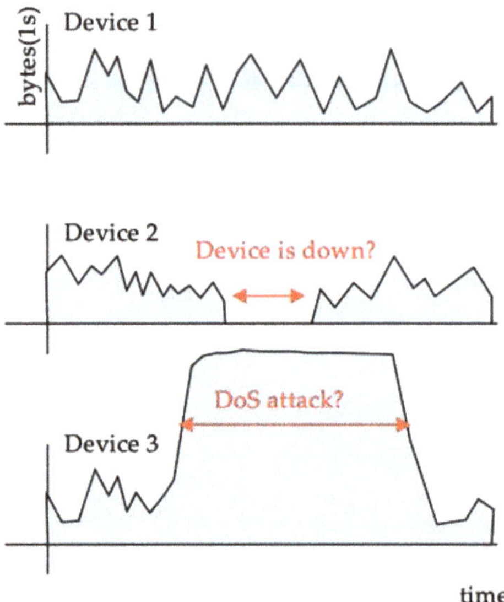

Figure 1. A disconnected device is detected using the traffic it generates.

Traffic time series can also be used for security monitoring. The analysis of anomalies in device traffic patterns can be used to detect erratic behaviour from the smart devices due to malfunctions or security violations [6]. Currently, the IoT ecosystem offers many devices with low security, and has already been used for the distributed denial-of-service (DDoS) attacks [7]. Monitoring the network activity from these devices to the centralised hosts (or to any other destination) is critically important for providing early intrusion detection.

Creating a per-device time series requires classifying each network packet when hundreds of thousands of classes are defined. Each packet could belong to only one class; therefore, it is considered in the time series for one device, or multiple time series could be created for each device. For example, all of the traffic that a device sends to its configured centralised collector could be recorded in a time series while separately accounting for all the traffic it sends to other destinations in a different time series (to detect anomalies). Therefore, time-series extraction requires the multi-label classification of packets, where a given packet may be assigned to several classes: one for each time series for which it has to be accounted.

Most sensors collect little and infrequent information; therefore, the network traffic they create presents a low bitrate. Low-rate information includes, for example, the location information sent from sensors installed in cars from a rental company. However, other sensors produce higher bitrates, for example, seismic sensors or surveillance cameras. In a traffic aggregation point or a location close to a collector in a large population of devices, millions of packets per second are expected [8], implying several gigabits of traffic per second.

Creating hundreds of thousands of time series based on source and destination when each packet can account for more than one class is not a trivial problem. This level of monitoring is typically performed by collecting NetFlow statistics from IP routers and postprocessing those flows to create the time series [9–11]. However, NetFlow monitoring presents time-resolution limitations. When a flow has finished, aggregated counters are provided. While the flow is active, the periodic dumps of all of the flows in memory of the networking device are sent to the NetFlow collector. Owing to the large amount of active flows, this collection has a periodicity of several minutes, thus losing all of the details below this scale. Most analysis and prediction algorithms require several measurement points in the time series, which would require several minutes to be collected from flow records. Reaction times in the order of several minutes are not appropriate for critical devices such as health monitoring sensors or probes in a nuclear plant. However, reducing the period of active flow collection in NetFlow statistics implies a higher load on the networking equipment, making it unfeasible.

The objective of this work is to validate that time-series extraction directly from a network packet stream can be performed sufficiently fast, and for a sufficiently large class set, to cope with the expected IoT scenarios. The traffic stream can be obtained by mirroring the packet flow at a network switch. This is a common functionality offered by most enterprise class switches. We focus on the extraction of traffic volume time series, namely the byte or packet counters, from a passive monitoring probe. Our test probe processes several gigabits per second of traffic in real time, providing the time series with a resolution better than one second.

Packet classification algorithms are a central part of this system. They assign each packet to a single class or multiple classes of equivalence. Packet classification is a well-studied problem [12–14]. If the number of classes is not large, a simple linear search for every packet over the class list is typically sufficient. However, depending on the traffic arrival rate and the computing power, above a certain number of classes, reviewing every class in sequence could consume too much time. Then, the analysis machine cannot cope with traffic at line rate without splitting the work among several central processing unit (CPU) cores, thereby increasing hardware costs. Several algorithms have been proposed and used to reduce the number of classes to visit [13,14]. Within hardware systems, optimal performance can be achieved using structures based on ternary content addressable memories (TCAMs) [15]. They can perform comparisons in parallel with all of their content cells. However, they incur high development costs for the ad-hoc hardware solution.

Software-based packet classification solutions sort the classes into binary search trees (also called "tries"). As classes typically involve IP sources and destination addresses, a hierarchy of search tries is used: first for the destination address, and subsequently for the source address. This hierarchy enables identifying the location of the relevant classes in only a few operations. The complexity in classification grows with the number of bits in the input classification information. However, this is at the expense of algorithm complexity and the memory that is required to store the search tries' structure. Probabilistic classification structures such as neural networks or Bloom filters, which are quite common in other classification scenarios, are not appropriate for the problem at hand. They have been used when high uncertainty in the classification is present or when the number of possible classes is very small [16,17]. On the contrary, the classification for time-series extraction in an IoT scenario is deterministic: it is based entirely on source and destination addresses, and it must classify packets when a large number of classes is defined. Neural networks and Bloom filters, being probabilistic, can produce wrong classification results. Failure in classification in one of these structures for a pair of source and destination network addresses does not happen for random independent packets, but rather for every packet in a time series between the same pair of addresses; therefore null or flat-line time series would result. On the other hand, both techniques require at least as many computing elements as the number of possible classes. For neural networks, at least one neuron is required for each possible class. Using Bloom filters, at least one filter is required for each class. Both structures can be implemented in parallel hardware architectures; however, in software implementations, they present a linear complexity with the number of classes, which in the present scenario we will show can reach hundreds of thousands of classes.

A survey of packet classification algorithms can be found in [12]. They are proposed for quality of service and security filtering scenarios. We herein evaluate how these algorithms apply to the time-series computation problem when the number of classes is as large as that expected in IoT scenarios.

To calculate the traffic time series for monitoring purposes, a packet has to be assigned to every class to which it belongs. The algorithm cannot stop whenever it obtains the best matching class, but it has to search over all the classes that may apply. We have selected three algorithms that can be used for multi-label classification: the linear list of rules search, the simple hierarchical tries [12], and the set-pruning tries [12]. Other popular algorithms such as the grid-of-tries [18] have to be discarded, because they do not comply with the multiclass requirement.

This work demonstrates that the hierarchical tries with recursive search can be used to build passive monitoring systems that can monitor tens of thousands of IoT devices in real time, with fast reconfiguration when devices are added or removed.

In Section 2 of this paper, the network scenario is presented, selected algorithms are described, and the methodology for performance evaluation is explained. In Section 3, the algorithms are compared using several performance metrics. Section 4 concludes the paper.

2. Methodology

2.1. Scenario

We consider an ISP offering network connectivity to IoT devices. These devices are separated into different groups. The network provider assigns addresses to the devices in order to create these groups. For example, a group could contain all of the devices in one building (or in the same region), or all of the devices with the same type of sensor. For some IoT applications (e.g., ubiquitous sensors), these groups may contain hundreds of thousands of devices, each one with an individual network address.

Network addresses are a sequence of bits that is used as a network locator, and are assigned according to the network protocol (IPv4, IPv6 ...). Different protocols use different numbers of bits in the sequence (32 in IPv4, 128 in IPv6...) or even variable length addresses. A given host, such as the controller for a group of devices, will be assigned a well-known network address. Groups of devices in the same network will be assigned an address in the same so-called subnetwork, namely a set of addresses with the same prefix bits. The first p bits in an address are called a prefix of size p.

Monitoring traffic from a large population of IoT devices requires classifying traffic coming from or going to different single network addresses or network address prefixes. Figure 2 shows the scenario under consideration. Several groups of devices communicate with different destinations through the ISP's network. This communication is bidirectional; the controllers may send requests to the devices, and the devices may send data to the controlling hubs. Monitoring this communication implies separating different views of the traffic. We can examine traffic from a single device or from a group of devices. We can be interested in traffic from a group of devices to one controlling server, or in the opposite direction from one controlling hub to a group of devices. It may be interesting to separate traffic from groups of devices to any address to detect whether devices are being used to perform denial-of-service attacks to Internet targets, or detect communications with suspected malware hubs.

An ISP monitoring system has to cope with this diversity of traffic flows. Every flow of interest can be described by a source and a destination. Every source and destination may be a single network address or a group of addresses given by a network address prefix with a length of p bits. We define a class as any combination of source and destination addresses in which the monitoring system is interested. A passive monitoring device that examines every network packet going through the ISP's network has to decide whether the packet belongs to any possible class of interest.

We denote $C = \{C_i\}$, where $C_i = (s_i, d_i)$ is the set of classes of interest for the monitoring system. The elements of C are our traffic-flow classes.

s_i is a source prefix, and d_i is a destination prefix. They may be prefixes, full addresses (a prefix of n bits), or the "any host" address (represented by a prefix with 0 bits).

S is the set of source prefixes of interest $S = \{s_i\}$. The number of elements in S is N_S. Similarly, D is the set of destination prefixes of interest $D = \{d_i\}$. The number of elements in D is N_D. C is the set of classes of interest. C is a subset of the Cartesian product $S x D$.

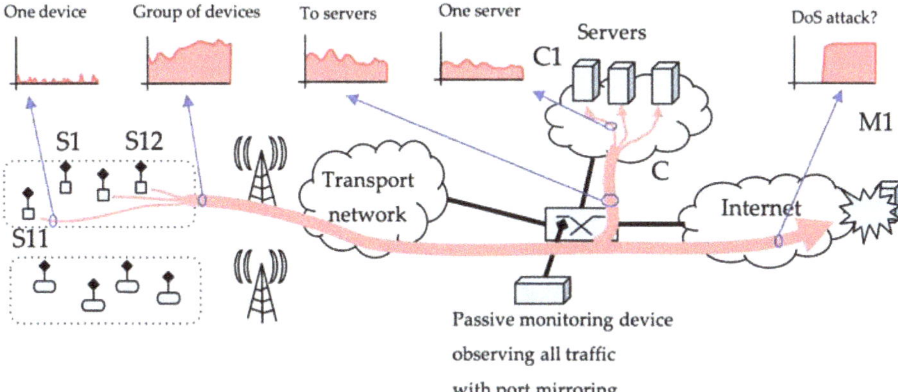

Figure 2. Network scenario with devices grouped by network address.

The number of elements of C that can be used may be as large as $N_S \times N_D$, in case we are interested in every possible source–destination combination. In a real scenario, only a subset of these combinations will have to be monitored, and the number of classes in the set will be $N_C < N_S N_D$.

As the number of addresses and prefixes of interest increases, it is more difficult for traffic-processing software to classify packets on a larger set of classes in real time.

2.2. Algorithms and Data Structures

We herein evaluate three popular algorithms for multi-label packet classification. The algorithms store the list of classes in different data structures. Every network packet can be matched against any class stored in these data structures. We compare the performance of these algorithms in scenarios with a large number of classes, N_C. We will measure the performance based on the packet processing speed and its memory footprint.

Table 1 contains the symbols describing the classes used in the data structure examples that follow. A packet may belong to several classes, i.e., a packet from S12 to C1 should be in classes one, three, and six.

Table 1. Class definitions for the examples in the algorithm and description of the data structures.

#class	Source	Destination	
1	S1	C1	Traffic from sensor group S1 to controller C1
2	S11	C1	Traffic from sensor S11 to controller C1
3	S12	C1	Traffic from sensor S12 to controller C1
4	S0	M1	Traffic from any sensor to malware host M1
5	S11	M1	Traffic from sensor S11 to malware host M1
6	S1	Any	Traffic from sensor group S1 to anywhere

A. Linear Search (LS) over the List of Classes

As a simple reference algorithm, using linear search (LS), every packet is evaluated sequentially against all of the N_C classes in C as shown in Figure 3. The list of classes is stored in a simple linked list or in an array. The expected per-packet processing time depends on the number of classes N_C as $O(N_C)$.

B. DStries with Recursive Search

A data structure called a DStrie [12] is built to store all of the classes. This structure is composed of several substructures called tries, which are created as follows.

Structure: List of classes for linear search

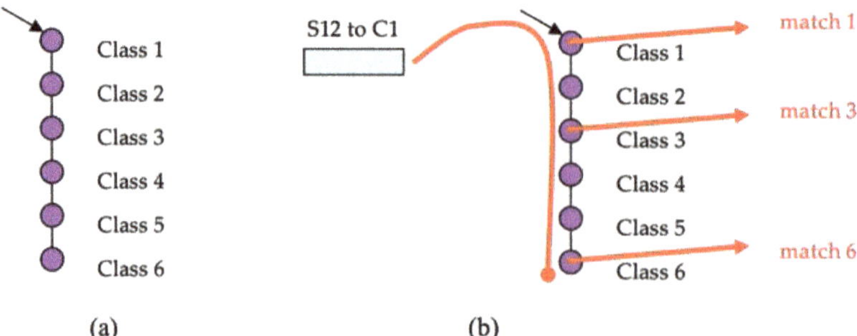

Figure 3. (**a**) Data structures and (**b**) search procedure in linear search (LS).

A binary search tree (trie) is a data structure that can store a set of network address prefixes. Prefixes are stored in a decision tree starting with the first bit of the address. For example, to store the binary prefix 0011, four nodes are created on the tree. The first bit is 0, the second is 0, the third is 1, and the fourth is 1. Data associated with prefix 0011 would be stored at the red node in Figure 4. Those associated with prefix 1001 would be stored at the green node in Figure 4, and those associated with the prefix of length 0 bits (representing any address) would be stored at the root node. Decision nodes that are not required to reach any of the stored prefixes are not in the tree. Every node in the tree is associated with a prefix, but that prefix may not be stored in the tree. For example, in Figure 4, the node for prefix 001 does not contain any stored information, but it exists because it is required to reach prefix 0011. Meanwhile, the node for prefix 010 is not in the tree. The tree is built by adding to all of the decision nodes that are required to reach the stored prefixes of the tree.

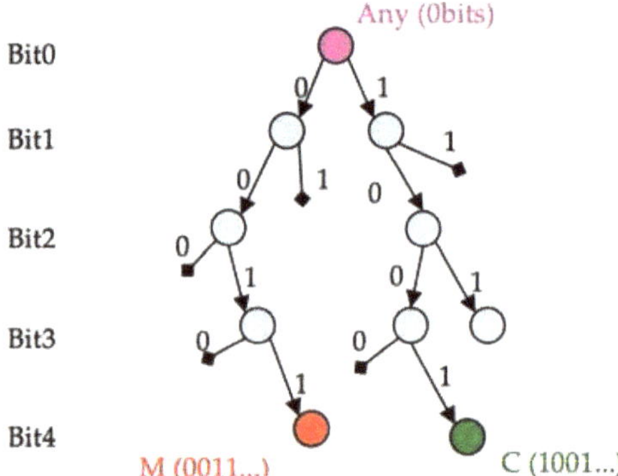

Figure 4. Destination prefix binary search tree (*trie*).

The DStrie data structure has to store classes with a source prefix and a destination prefix. It is composed of two levels of tries. The high-level trie stores destination prefixes. Every time a packet is examined, its destination address is searched for in the destination trie to find all of the destination prefixes that may apply to that address. A class also contains a source prefix; therefore, every node in the destination trie that stores a prefix also stores a pointer to a second-level trie containing the source prefixes.

A given traffic class is stored in the node that is indicated by its source prefix in the source trie that is pointed to by the node in the destination trie corresponding to its destination prefix. An example is shown in Figure 5.

Structure: DStries for Recursive Search

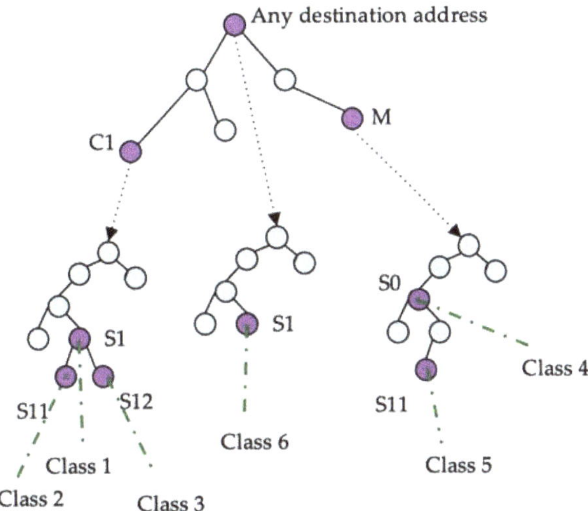

Figure 5. DStrie for recursive search example.

To classify a packet with a given source and destination address, first, the destination address is searched for in the destination trie. After the destination is found, the source is searched for in the corresponding source trie. A packet that verifies a destination node also verifies every destination in the path from the root to the final node in the destination trie. Thus, a given packet has to be searched for in every source tree of any destination node in the path, from the root to the best destination prefix.

The classes are stored in only one place in the structure; however, to search for every possible class, multiple source tries must be searched recursively. In the example shown in Figure 6, to classify a packet with source S12 and destination C1, the destination address is searched for in the topmost trie. Two nodes match: the root node corresponding to any address, and the node corresponding to C1. In both nodes, the source trie searched for the source address. In the first one, node S1 points to class six. In the latest, the search in the source trie obtains S1 and S12 pointing to classes three and one, respectively. Thus, the search results in three hits.

The number of operations that is required to match an address in this type of trie depends on the average bit length of the prefixes that are stored in the trie, as this yields the average depth of the binary tree. In our evaluation scenario, a significant amount of full-length prefixes are stored in the destination trie (individual devices), as well as in every source trie. Therefore, the expected time to match an address down a tree should be $O(n)$, where n is the number of bits in network addresses.

The time to process a packet is given by the time that it takes to search a source address in a source tree multiplied by the average number of source tries that have to be traversed, which is also $O(n)$. Therefore, the total time should be in the order of $O(n2)$.

C. Set Pruning DStries

This algorithm presents improvements over DStries with recursive search. A trie is built with every possible destination prefix as before. However, every class is stored not only in the source tree given by its destination, but also in the source tries of its children in the destination trie. A class is stored multiple times, thereby increasing the amount of memory that is used by the data structure. A certain node in a source trie may point to a list of classes instead of a single class. In Figure 7, the source trie pointed to by node C1 contains several classes at node S1.

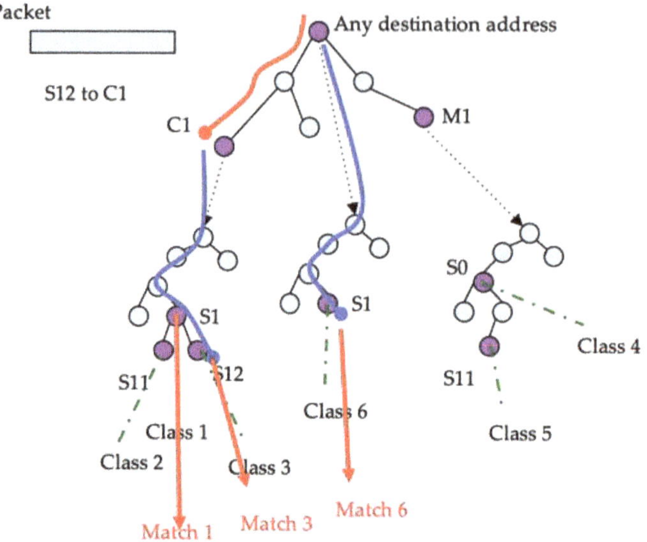

Figure 6. Search procedure example traversing a DStrie.

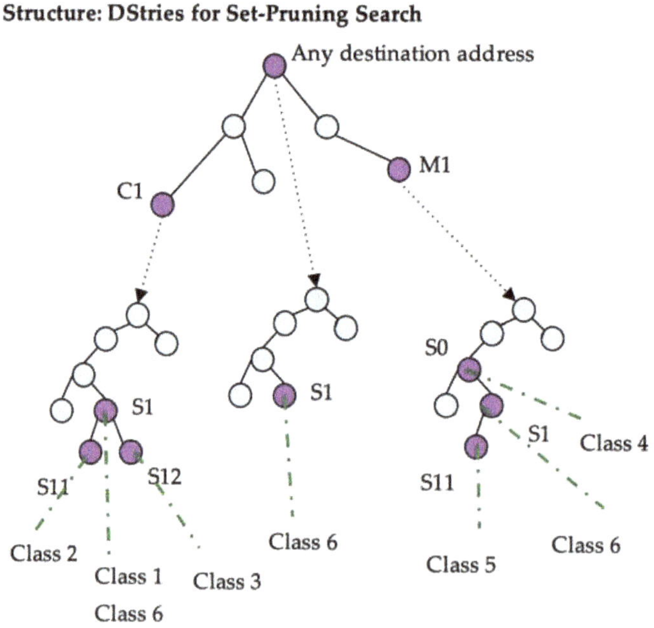

Structure: DStries for Set-Pruning Search

Figure 7. Set-pruning DStries example.

In this algorithm, the search that is undertaken for every packet is simpler than that in DStries with recursive search (see Figure 8). Every packet destination address is searched for the best destination match. Afterwards, the source address is searched for through a single source tree for all of the matches. A single source trie is inspected per packet, but typically, the source tree is denser. The algorithm consumes more memory by storing redundant information to achieve a faster structure traversal.

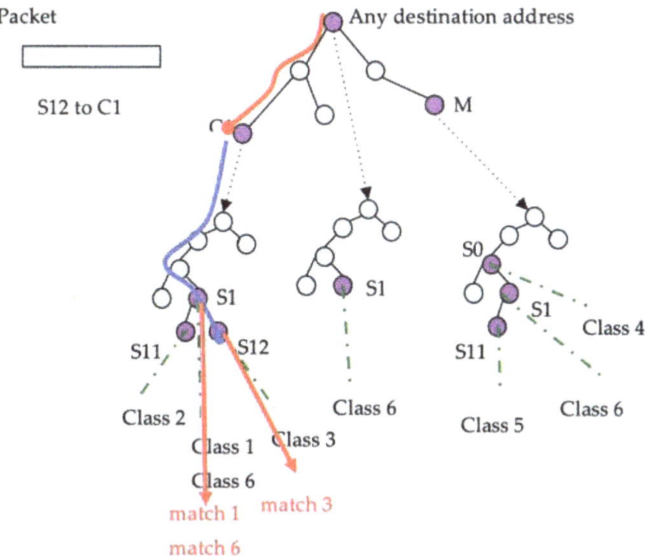

Figure 8. Search example traversing a set-pruning DStrie.

The expected time to match a packet is in the order of traversing a search tree for the destination address, and subsequently another one for the source address. Therefore, the expected running time is $O(2n)$.

We have implemented all of the algorithms mentioned herein inside a custom time-series computing software that uses a pcap [19] file as input. We use it to evaluate the computing time for several real-world traffic traces.

2.3. Network Traffic for Performance Evaluation

To evaluate the performance of these algorithms, network traffic and sets of classes are required. Several traffic captures have been used to provide network packets for classifications (Table 2). The class list to observe is generated from the prefixes and addresses that are present in the captured traffic. The purpose is to generate a realistic scenario. If random prefixes and addresses were generated, most of them would be discarded quickly in the classifier, because they could be found in short branches of the tries. This would yield a false positive bias towards the DStries methods.

We extract every network address that is present in the traffic trace, as well as every network prefix using prefix length $p = 24$. This list is used to generate every possible combination of a source prefix and destination address, and its corresponding source address and destination prefix. This set of combinations is used as the maximal classes set. To test an algorithm, a subset of N_C of these classes is randomly extracted. The trace is processed with this set of classes to evaluate the processing speed. The average number of packets that is processed per second is recorded, as well as the time spent building the classes' structure into memory and the total memory footprint.

For every classification algorithm, the experiment is repeated using random subsets of classes with different N_C sizes.

Table 2 shows the primary statistics from the traffic traces that are used in the evaluation. All of them are IPv4 ($n = 32$ **bits**) packet traces. Trace upna1h was captured at a university Internet access link supporting the traffic from more than 2000 devices. Trace iot1h was captured at a production IoT network with at least 200,000 operating devices. We captured approximately one hour of traffic at both links. Each of the traces contains more than 50 million packets. When real IoT traffic is not available, the research community resorts to generic IP traffic from desktop computers. Using trace upna1h, we show that the absolute values in the evaluation depend heavily on the traffic pattern. No valid conclusions could have been extrapolated from generic Internet traffic alone. The specific case of traffic from a real IoT scenario must be taken into consideration in the evaluation, as we do.

Table 2. Global statistics from the evaluation traffic traces.

trace	duration	#packets	bytes	throughput	average packet size	#unique address	#prefixes 24 bits	#classes
upna1h	1 h	193 M	170 GB	381 Mbps	~880 B	290 k	211 k	1193 k
iot1h	1 h	58 M	9 GB	24 Mbps	~165 B	214 k	8.9 k	724 k

Both traces contain a large number of different network addresses in use (more than 200,000). Trace iot1h concentrates them into approximately 9000 prefixes ($p = 24$) while upna1h spreads them out over more than 200,000 prefixes. For upna1h, a maximal class set is built, containing 1.19 million classes. For iot1h, the maximal class set contains more than 724,000 classes.

Figure 9 shows the different experimental cumulative distribution functions of packet sizes in each of the traffic traces. The average packet size (indicated by the dashed red line) is much larger in a generic Internet access scenario (trace upna1h) than in a real IoT scenario (trace iot1h). These sizes will affect the packet processing speed, as will be shown in the next section.

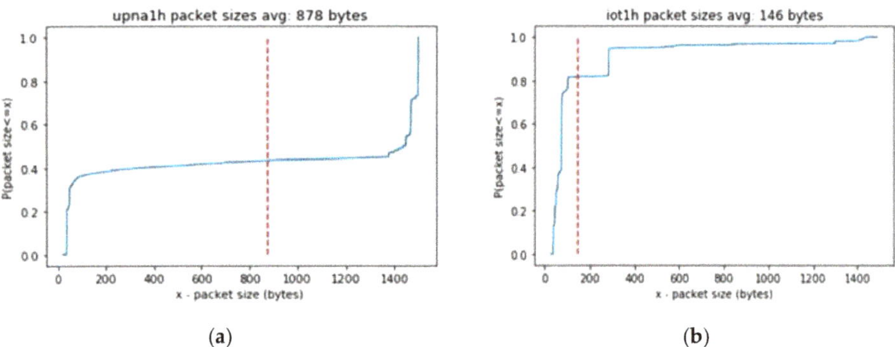

(a) (b)

Figure 9. Cumulative distribution functions of packet sizes (**a**) in upna1h traffic trace, or (**b**) in iot1h traffic trace.

Traffic processing is performed using a single core in a Xeon E5-2609 CPU at 1.7 GHz with 128 GB of random access memory (RAM). The trace is preloaded in RAM to measure the packet processing time without disk access influence.

3. Results and Discussion

The input traffic traces were processed by the described algorithms using different random sets of N_C classes. Figure 10a shows a two-minute fragment of the time series for the total traffic in trace upna1h. Figure 10b shows the extracted time series for one of the classes. In this section, we evaluate the traffic intensity that a network probe could withstand to create several time series for thousands of IoT devices in real time.

We span a range from hundreds to hundreds of thousands of classes in the set (i.e., number of time series to compute simultaneously). The number of supported devices depends on the number of classes (number of time series) per device. These time series are computed from the input traces using the three presented algorithms. Sets containing up to 5000 classes are evaluated using the three algorithms. Larger sets are simulated using only the fastest methods. The throughput obtained in packets per second, using up to 5000 classes, is shown in Figure 11a (for upna1 trace) and Figure 11b (for iot1h trace).

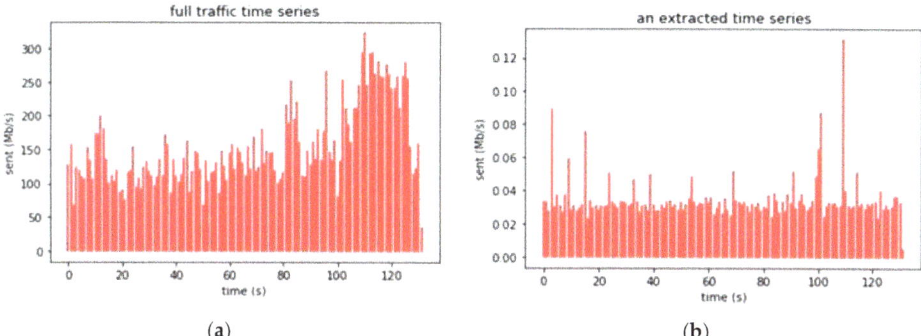

Figure 10. Traffic time series for two minutes (**a**) in the whole trace upna1h, or (**b**) in only one user in trace upna1h.

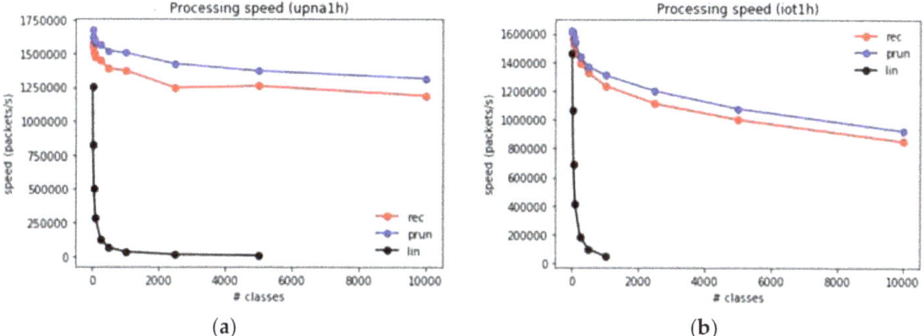

Figure 11. Packet processing speed for small sets of classes. (**a**) upna1h trace (**b**) iot1h trace.

The throughput depends strongly on N_C. For the linear search algorithm, the computation time increases as $O(N_C)$; therefore, the throughput decreases proportional to $\frac{1}{N_C}$. This is an extremely fast decay compared to the other algorithms, thereby causing the LS to be non-competitive for thousands or more classes.

Recursive search and set-pruning DStries algorithms decay more slowly, and achieve more than one million packets/s for thousands of classes.

The experiment was extended to at least 724,000 classes, excluding the computation of the LS algorithm, whose results are predicted easily. Figure 12a,b show the packet processing speed for both traces.

Both algorithms present a similar behaviour in the number of processed packets per second, with a slightly higher performance in the case of set-pruning DStries. The throughput in gigabytes (Gb)/s depends on the average packet size. Figure 13 shows the achievable processing bit rate when considering the average packet sizes. The average per packet processing time in our reference probe is about four microseconds when 100,000 classes are used, and less than 18 microseconds for 700,000 classes. Less than 1% of the packets suffer a processing time larger than 200 microseconds, even when 700,000 classes are used. Therefore, the error margin in time-series computation with samples every second is low.

The upna1h traffic trace presents average Internet packet sizes; therefore, the multigigabit per-second processing speeds are reached. Trace iot1h was obtained from a real IoT scenario, where the packets were smaller than the average sizes in the Internet. For a similar number of processed packets per second, a much smaller traffic bitrate was consumed (approximately one Gb/s). We found it important to consider the specific traffic characteristics in real IoT traffic.

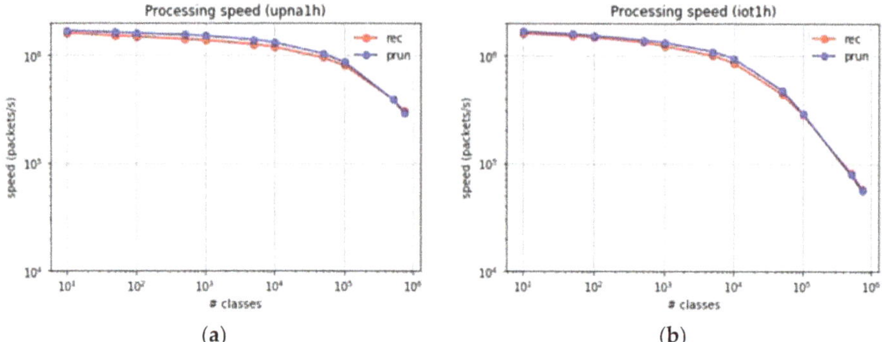

Figure 12. Packet processing speed for large sets of classes. (**a**) upna1h trace. (**b**) iot1h trace.

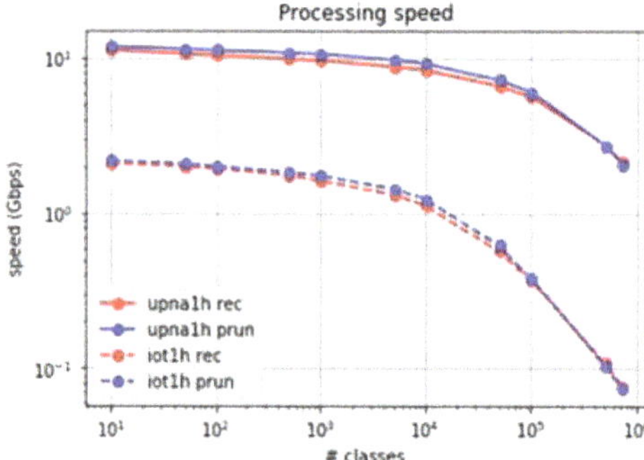

Figure 13. Processing speed in bits per second comparison for upna1h and iot1h traces.

From these results, we can estimate the number of IoT devices that a single-core CPU could analyse. The limit depends on the traffic behaviour; therefore, we have included results based on a typical IoT profile (extracted from trace iot1h), and on a generic Internet access profile (extracted from trace upna1h).

A single point in Figure 12 represents the maximum average packet arrival rate that could be processed for a specific number of classes, N_C. This number of classes is simply the number of IoT devices multiplied by the number of time series per device. The maximum packet arrival rate corresponds to the aggregate of traffic from the same number of devices. Assuming certain traffic intensity or an average per-device packet arrival rate, the maximum number of devices can be obtained. We introduce this intensity as a parameter using the average number of packets per second sent by each device. For example, a sensor that collects temperature measures and contacts its central repository once per minute, sending all of the measures with the exchange of 12 network packets generating $12/60 = 0.2$ packets per second (pkt/s).

Figure 14 shows the maximum number of devices supported in real time versus the number of time series to compute per device. As a bidirectional time series requires two classes, the case of six time series (six classes) corresponds to three bidirectional time series. This could represent, for example, the case of an operator who wants to measure the bidirectional traffic from each IoT device to its central collector, any server in its server farm, and everywhere else. The results for 10 time series consider the possibility of two extra bidirectional time series per device. We have included the results for up to 20 classes per device, or 10 bidirectional time series. Figure 14 is based on the devices that generate an average of five pkt/s.

The results depend on the traffic profile, or how the random traffic process is generated. For a present IoT network scenario with six classes per device, and an average of five pkt/s per device, more than 40,000 devices are supported. Using 20 classes, at least 20,000 devices are supported in a single-core CPU processing machine. For a traffic profile similar to a generic Internet access link, better results are obtained, especially for low numbers of classes. The results for typical IoT scenarios are worse, owing to the small average packet sizes.

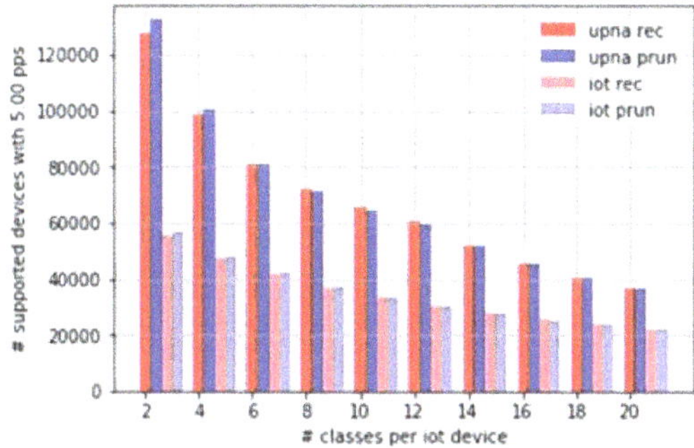

Figure 14. Number of supported devices that generate an average of five pkt/s.

Figure 15a,b show the number of supported devices when the amount of traffic generated by each device is increased from five pkt/s up to 40 pkt/s. Figure 15a considers six time series per device, and Figure 15b shows the requirement of 10 time series per device. This represents future scenarios where IoT devices collect more frequent measurements or send larger files to the central collector. For example, a camera creating a 15-kB image every second could easily result in a traffic intensity of 15 or 20 pkt/s, considering both directions of the traffic.

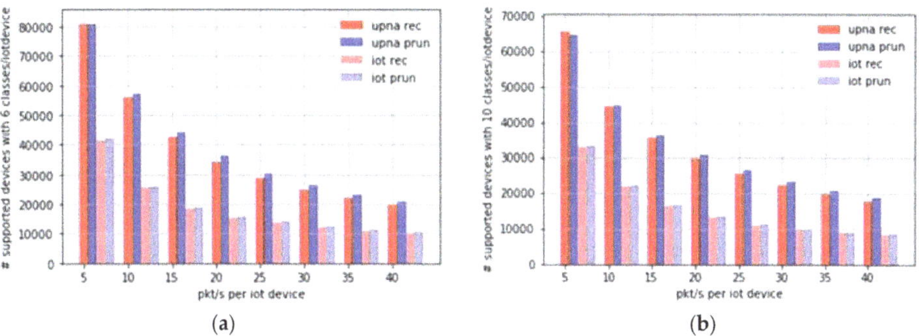

Figure 15. Number of devices supported, depending on the traffic generated per IoT device (**a**) extracting 6 time series per device, or (**b**) extracting 10 time series per device.

Figure 15 shows that by using six classes (three bidirectional time series) and high traffic intensity-generating devices (up to 40 pkt/s), at least 10,000 devices would be supported by both algorithms.

Although both set-pruning methods (DStries and DStries with recursive search) offer similar speed results, they present some hidden drawbacks that must be considered. The data structure that set-pruning DStries stores in memory duplicates the class information; therefore, its memory footprint is higher than the one from a simple recursive search or the linear search method. Figure 16 shows the memory usage for both DStries-based algorithms and both traffic traces. Although the linear method presents minimal memory usage, it must be discarded as a suitable algorithm, owing to its low performance in terms of packets processed per second for a large number of classes.

The second drawback in set-pruning DStries is the difficulty in updating its memory structures. Whenever a new class has to be added to the structure, the whole structure must be recreated from scratch. This is simpler than updating the structure incrementally. Adding (or removing) classes must be performed whenever new IoT devices are added to the network and monitoring platform. This is typical in large deployment scenarios, as it could be in household metering devices in an electric company. Figure 17 shows the time that is required to build the DStrie in the reference computer when a single new class must be added to an already existing structure. The time to build a set-pruning DStrie for hundreds of thousands of classes may reach several minutes; thus, the algorithm is not useful if the set of classes to monitor must change frequently.

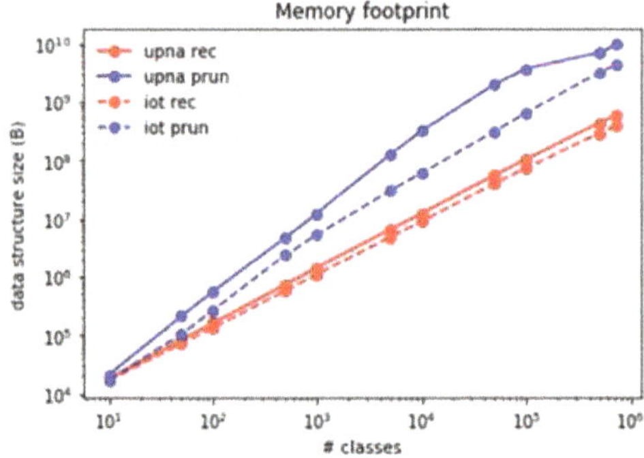

Figure 16. Comparison of memory usage.

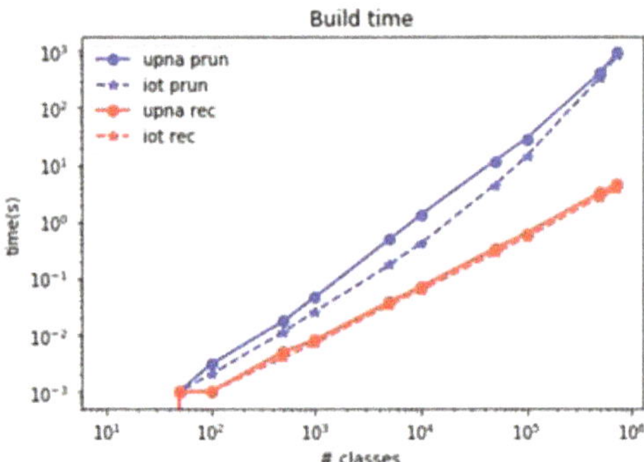

Figure 17. Comparison of required time to add a new class.

4. Conclusions

We herein demonstrated how a single-core CPU could create the traffic time series for tens of thousands of IoT devices, even when several time series were required for each device. The results depended on the number of time series desired, and the traffic intensity created by each IoT device. Considering different traffic profiles, the results were validated using traffic from a real IoT deployment scenario with more than 200,000 devices, and from a generic Internet access link.

The algorithms of linear search, DStries with recursive search, and set-pruning DStries, were evaluated. Although the linear search algorithm presented the simplest implementation and the lowest memory requirements, it provided the worst results regarding the number of time series that it could compute in real time compared to the other algorithms. It could only be used in extremely small scenarios. Both DStries with recursive search and set-pruning DStries yielded similar results in computation speed, and hence the number of devices supported in real time. DStries with recursive search exhibited lower memory requirements compared to set-pruning DStries; therefore, it is better suited for scenarios with low computing power and memory available in each time-series computing node. Set-pruning DStries also created complex in-memory structures, and required more time when a new IoT device had to be added or removed. For highly dynamic scenarios where new nodes are added frequently, DStries with recursive search is the most suitable algorithm, as it offers the lowest memory footprint and the lowest modification time. In a single-core CPU, it can create three bidirectional time series for each one of more than 20,000 IoT devices in real time, when each device sends an average of 10 packets per second.

Author Contributions: Conceptualisation, M.I., D.M. and E.M.; Data curation, M.I., D.M. and S.G.-J.; Formal analysis, M.I., D.M., E.M. and S.G.-J.; Methodology, M.I., D.M. and E.M.; Software, M.I. and S.G.-J.; Validation, M.I., D.M. and S.G.-J.; Writing—original draft, M.I.; Writing—review & editing, M.I., D.M. and E.M.

Funding: This work is funded by Spanish MINECO through project PIT (TEC2015-69417-C2-2-R).

Conflicts of Interest: The authors declare no conflict of interest.

References

1. Al-Fuqaha, A.; Guizani, M.; Mohammadi, M.; Aledhari, M.; Ayyash, M. Internet of Things: A Survey on Enabling Technologies, Protocols, and Applications. *IEEE Commun. Surv. Tutor.* **2015**, *17*, 2347–2376. [CrossRef]
2. Guo, H.; Heidemann, J. IP-Based IoT Device Detection. In Proceedings of the 2018 Workshop on IoT Security and Privacy—IoT S&P '18, Budapest, Hungary, 20–25 August 2018.
3. Perera, C.; Zaslavsky, A.; Christen, P.; Georgakopoulos, D. Sensing as a service model for smart cities supported by Internet of Things. *Trans. Emerg. Telecommun. Technol.* **2014**. [CrossRef]
4. Marwat, S.; Mehmood, Y.; Khan, A.; Ahmed, S.; Hafeez, A.; Kamal, T.; Khan, A. Method for Handling Massive IoT Traffic in 5G Networks. *Sensors* **2018**, *18*, 3966. [CrossRef] [PubMed]
5. Shafiq, M.Z.; Ji, L.; Liu, A.X.; Pang, J.; Wang, J. Large-scale measurement and characterization of cellular machine-to-machine traffic. *IEEE/ACM Trans. Netw.* **2013**. [CrossRef]
6. Lakhina, A.; Crovella, M.; Diot, C. Characterization of network-wide anomalies in traffic flows. In Proceedings of the 4th ACM SIGCOMM Conference on Internet Measurement—IMC '04, Sicily, Italy, 25–27 October 2004.
7. Kolias, C.; Kambourakis, G.; Stavrou, A.; Voas, J. DDoS in the IoT: Mirai and other botnets. *Computer* **2017**. [CrossRef]
8. Kuang, J.; Waddington, D.G.; Lin, C. Techniques for fast and scalable time series traffic generation. In Proceedings of the 2015 IEEE International Conference on Big Data, Santa Clara, CA, USA, 29 October–1 November 2015.
9. Bull, P.; Austin, R.; Popov, E.; Sharma, M.; Watson, R. Flow based security for IoT devices using an SDN gateway. In Proceedings of the 2016 IEEE 4th International Conference on Future Internet of Things and Cloud (FiCloud), Vienna, Austria, 22–24 August 2016; pp. 157–163.
10. Miguel, M.L.F.; Penna, M.C.; Nievola, J.C.; Pellenz, M.E. New models for long-term Internet traffic forecasting using artificial neural networks and flow based information. In Proceedings of the 2012 IEEE Network Operations and Management Symposium, Maui, HI, USA, 16–20 April 2012.
11. Malott, L.; Chellappan, S. Investigating the fractal nature of individual user netflow data. In Proceedings of the 2014 23rd International Conference on Computer Communication and Networks (ICCCN), Shanghai, China, 4–7 August 2014.
12. Gupta, P.; McKeown, N. Algorithms for packet classification. *IEEE Netw.* **2001**, *15*, 24–32. [CrossRef]
13. Lim, H.; Lee, S.; Swartzlander, E.E. A new hierarchical packet classification algorithm. *Comput. Netw.* **2012**, *56*, 3010–3022. [CrossRef]
14. Liu, Z.; Sun, S.; Zhu, H.; Gao, J.; Li, J. BitCuts: A fast packet classification algorithm using bit-level cutting. *Comput. Commun.* **2017**, *109*, 38–52. [CrossRef]
15. Orosz, P.; Tóthfalusi, T.; Varga, P. FPGA-Assisted DPI Systems: 100 Gbit/s and Beyond. *IEEE Commun. Surv. Tutor.* **2018**, *1*. [CrossRef]
16. Bloom, B.H. Space/Time Trade-offs in Hash Coding with Allowable Errors. *Commun. ACM* **1970**, *13*, 422–426. [CrossRef]

17. Turčaník, M. Packet filtering by artificial neural network. In Proceedings of the International Conference on Military Technologies (ICMT), Brno, Czech Republic, 19–21 May 2015; pp. 1–4.
18. Srinivasan, V.; Varghese, G.; Suri, S.; Waldvogel, M. Fast and scalable layer four switching. *ACM SIGCOMM Comput. Commun. Rev.* **1998**, *28*, 191–202. [CrossRef]
19. Packet Capture Library LIBPCAP. Available online: http://www.tcpdump.org (accessed on 24 October 2018).

Article

A Feature-Based Model for the Identification of Electrical Devices in Smart Environments

Andrea Tundis [1,*], Ali Faizan [2] and Max Mühlhäuser [1]

[1]   Department of Computer Science, Technische Universität Darmstadt, Hochschulstrasse 10, 64289 Darmstadt, Germany; max@tk.tu-darmstadt.de
[2]   Software AG, Uhlandstraße 12, 64297 Darmstadt, Germany; ali.faizan@softwareag.com
*   Correspondence: tundis@tk.tu-darmstadt.de; Tel.: +49-6151-16-23199

Received: 5 April 2019; Accepted: 5 June 2019; Published: 8 June 2019

Abstract: Smart Homes (SHs) represent the human side of a Smart Grid (SG). Data mining and analysis of energy data of electrical devices in SHs, e.g., for the dynamic load management, is of fundamental importance for the decision-making process of energy management both from the consumer perspective by saving money and also in terms of energy redistribution and reduction of the carbon dioxide emission, by knowing how the energy demand of a building is composed in the SG. Advanced monitoring and control mechanisms are necessary to deal with the identification of appliances. In this paper, a model for their automatic identification is proposed. It is based on a set of 19 features that are extracted by analyzing energy consumption, time usage and location from a set of device profiles. Then, machine learning approaches are employed by experimenting different classifiers based on such model for the identification of appliances and, finally, an analysis on the feature importance is provided.

Keywords: electrical devices; classification; energy management; machine learning; smart environment

---

1. Introduction

Through the integration and interconnection of software-centered devices, traditional power system, which are typically centered on mechanical and electrical components, are supported by more complex equipment which enable more advanced functionalities in terms of management and control. The result of such evolution which makes those systems more intelligent is named as Smart Grid (SG) [1,2]. The advantage of having software components in the network enables the introduction of more sophisticated mechanisms for monitoring the SG as well as to support, in a more effective way, the decision process for the dynamic energy distribution according to the current use of energy, resource state and weather conditions [3–6]. Additionally, from the consumer perspective, specific optimization techniques can be exploited for managing the scheduling of the device usage (for example based on specific hours of the day or week according to the electricity costs) in order to save money. Moreover, as an actor can be producer and consumer (so called prosumer) of electricity in a SG [7], such flexibility enables a different way to distribute the energy and to deal with unexpected emergency situations, resulting from faults and failures in the network [8–10]. A general overview of a Smart Grid is depicted in Figure 1, which includes City and Buildings, Power Plants, Wind Turbines, Electric Vehicles, Solar Panels and Smart Homes.

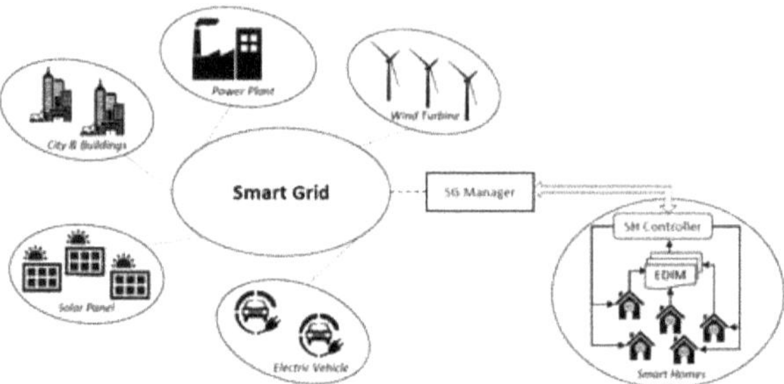

Figure 1. Smart grid overview.

In this scenario, an important role is played by Smart Homes (SHs). They are equipped with electrical devices that can be controlled and monitored remotely not only to achieve economic benefits by saving electricity, but also by contributing in the reduction of carbon dioxide emission in the environment [11]. SHs represent the human side of the SG [12]. They provide a new perspective towards the usage of the energy in the everyday life and, in particular, in the relationship between energy utilities and consumers. Typically, the traditional homes have devices that work locally and manually, usually by switching them on/off by pushing a button, with a limited control in terms of their automatic management. A SH, instead, represents the convergence of energy efficient, controllable electrical appliances and real-time access to energy usage data. This combination of device management and smart grid enables proactively managing energy use in ways that are convenient, cost effective, and good for the environment.

To enable such flexibility, the envisioned communication mechanism foresees two main components: (a) a *Smart Grid Manager (SG Manager)*, which has a global view of the network. It is responsible for making decisions about the energy distribution on the basis of the overall available resources; (b) a *Smart Home Controller (SH Controller)*, which represents an interface between the SG and a house. Every *SH Controller* aims, from one side, to retrieve information regarding the electricity consumption in a house and provides it to the *SG Manager*, and from the other side to manage the electrical devices in the house on the basis of habits and rules specified by the user.

Behind the advantages of a more intelligent energy grid management, one of the main challenges for enabling such a pro-active control relies on the automatic recognition, identification and classification of the electrical appliances. This in turn requires facing several factors [13], such as: (i) *power consumption extraction* that is the process of measuring the energy from different devices in order to identify recurrent consumption patterns; (ii) *multi-mode functionality*, this means that some devices can have multiple operation mode which can be misleading for their identification due to such a complex behavior; (iii) *parallel usage*, this is an important factor that has to be faced, since typically more than one device is in operation at the same time; (iv) *similar characteristics* because many devices can present similarities in the way they use the energy (e.g., consumption, charging time); (v) *external effects* because the data could be spoiled by external and random factors, which are not predictable, such as temperature, communication failures, human influences, etc.

In this context, this work proposes a model for the automatic identification of electrical devices, after they are plugged into an electrical socket. It is the basis for the automatic control of whole functionality of an SH which includes applications regarding, for example, the monitoring and control of appliances, the dynamic load management system based on available resources, power saving by scheduling the devices as well as emergency systems in case of faults or failures of specific resources in the network. The model is based on a set of 19 features which are able to characterize different electrical devices and distinguish them from others. They are derived by analyzing three main aspects: (i) *power consumption*:

related to the electricity being consumed by a device for a certain period of time, (ii) *working schedule*: which includes the hours of the days and the time duration when the device is turned on/off, and (iii) *location*: which represents the place where an electrical device is connected on the basis of the electrical socket within the house. Then, machine learning techniques are used to experiment the model through different classifiers, by using a dataset of 33 types of appliances [14]. The goodness of the proposed model is evaluated in terms of its accuracy, for the identification and classification of electrical devices, with respect to the existing approaches discussed in Section 2.

The rest of the paper is structured as follows. In Section 2, the related works about automatic devices identification are presented, whereas a background on machine learning techniques is reported in Section 3. Then, the adopted research approach is described in Section 4, whereas the proposed model as well as the features are elaborated in Section 5. The experimental results of the classification are highlighted and discussed in Section 6, whereas Section 7 concludes the paper.

2. Related Work

This section discusses the most relevant research efforts and related solutions, which have been proposed for supporting the identification of electrical devices.

In particular, in [15], a middle layer to connect sockets and devices, which is centered on Measurement and Actuation Units (MAUs), is presented. The MAUs monitor and analyze the electrical power consumption of any connected device individually by providing fine grain analysis. The main information for the classification is based on temporal behavior of the appliances, power consumption, shape of the power consumption, and level of noise. Different classifiers have been experimented with, but better performances have been reached by the Random Forest, LogitBoost, Bagging and the Random Committee, which achieved 95.5% accuracy.

In [13], different energy measurements, such as active power, reactive power, phase shift, root mean square voltage and current, by collecting data of each device in an isolated way, are instead considered. This approach aims to provide a plug and play tool to create energy awareness on the basis of real-time energy consumption of electrical devices. Additionally, multi-mode functionality, parallel usage of devices and external effects are also tackled. The difficulties to support the identification of devices which have multi-mode operation compared to those with a single operation mode are discussed. It resulted in an extensive training by deriving a classification model with an accuracy between 94–97%.

In [16], an approach centered on plug-based low-end sensors for measuring the electric consumption at low frequency, typically every 10 s, is presented. In particular, a sensor called PLOGG is used to record a vector of electrical parameters related to the appliance being monitored [17]. However, for each appliance class, a stochastic model is built from the observed consumption profiles of several instances of each class that are used to train the models. A k-NN classification algorithm has been employed on the basis of the identified features by reaching a level of accuracy equal to 85%. However, in [18], a plug and play "smart plug" is investigated. It aims to recognize the consumer appliance category, which is specified according to consumption scales and priorities, based on the employment of specific sensors. It allows for measuring and recording instantaneous energy consumption, by estimating specific parameters of consumer appliances such as the total harmonic distortion and the power factor.

In [19], an approach based on Non Intrusive Appliance Load Monitoring (NIALM) at meter level, to detect whether the device is switched on or off, is discussed. When a change occurs in the overall electrical power signal of the house, the change is analyzed and compared to the already-known patterns available in a database. Another centralized approach, for monitoring power signal, exploits the ZigBee device, which is attached to the main electrical unit [20]. It is used to identify in real time the appliance that contributes to each spike of energy. Another research effort, based on a centralized approach, is described in [21], in which the authors used a custom data collector and, in particular, a power interface oscilloscope and a computer as hardware. It allows for detecting electrical noise to classify electrical devices in homes by exploiting the electrical noise as an additional parameter. However, time series measurements, which represent electrical signatures of different electrical devices, are used in [22] for their identification.

A summary of the above-mentioned related works is reported in Table 1. Two main approaches, to face with automatic identification of electrical devices, emerged from the above related works. One is based on the employment of additional monitoring devices either distributed [15,22] or centralized [19–21] which results expensive in terms of money for their installation and hardly scalable; the second one that does not exploit any additional devices, is centered on energy measurements [13], but it lacks in the categorization and formalization of the adopted features. Some of those works used aggregated traces (AT) of multiple devices and attempt to disaggregate energy usage, whereas other works, as in our case, used directly disaggregated traces (DT). For the sake of the completeness of this paper, Table 1 provides a high-level overview of prior works by highlighting, for example, the used parameters, data collection techniques in terms of type of traces and accuracy they achieved.

Table 1. Comparison of the related works.

Related Work	Main Parameters	Additional Devices	Accuracy (%)	Adopted Approach	Trace Type
[13]	Active and Reactive Power Phase shift, V_{rms}, I_{rms}	None	94–97	Not specified	Not specified
[15]	Power consumption, Working schedule	Measurement and Actuation Units	95.5	Distributed	DT
[16]	Power consumption at low frequency	Plug-based low-end sensor	85	Distributed	DT
[19]	Power consumption	NIALM device	Not reported	Centralized	AT
[18]	Power factor Harmonic distortion	Smart Plug	Not reported	Distributed	DT
[20]	Active power, Reactive power, Phase shift, Signature length, Root mean square voltage, Sampling frequency	Zigbee Monitor	95	Centralized	AT
[21]	Electrical Noise	Oscilloscope, Laptop Custom Data Collector	85–90	Centralized	AT
[22]	Active power, Reactive power, Root mean square voltage, Phase shift	Smart Plug	93.6	Distributed	DT

Other proposals are available in literature, which are not reported and compared in Table 1 because some of them are based on different approaches and/or input data whereas others have different purposes. For example, there have been other works in the context of appliance identification that are centered on both different approaches and input data, such as those based on high frequency conducted electromagnetic interference (EMI) which use Non-Intrusive Load Monitoring (NILM), as described in [23]. It aimed to present some of the key challenges towards exploiting EMI and the dataset of the collected data, which was used in the experiment, is also available online and freely downloadable [24]. However, the work in [25] proposes a technique that aims at identifying anomalous appliances in buildings by using aggregate smart meter data and contextual information in near real time.

In this wide context, our paper is strictly related to those works, reported in Table 1, which dealt with disaggregated traces. In particular, our work stands out from the previous ones because (i) a set of features that characterize electrical devices are proposed and formalizedl (ii) a model, based on their combination, is used to identify and recognize devices when they are plugged into the circuit without additional monitoring devices and based on disaggregated traces, and (iii) high performances in terms of accuracy are reached.

3. Background on Machine Learning

Machine learning (ML) is a data analysis technique, based on computational algorithms able to learn directly from the data without a predefined model [26]. In particular, ML techniques aim to identify patterns from which the extracted information is used to make better forecasts, prediction and decisions.

Thanks to the huge amount of available data, ML represents a popular approach that is exploited in several fields, for facing classification-related problems, such as in (i) computational finance for the evaluation of credit risk and algorithmic trading; (ii) image processing and artificial vision for facial recognition, motion detection and object identification; (iii) computational biology for the diagnosis of tumors, pharmaceutical research and DNA sequencing; (iv) energy production for price and load forecasts; (v) automotive, aerospace and manufacturing sectors, for predictive maintenance; and (vi) natural language processing, for speech recognition applications.

Among the variety of existing techniques, an overview of the most popular ones is provided and briefly discussed below, such as Decision Trees (DTs), Support Vector Machine (SVMs), Linear Regressions (LRs), Naive Bayes (NB), Random Forest (RF), Random Committee (RC), Boosting, Bagging, and Artificial Neural Networks (ANNs). Some of them are then selected and taken into account in the evaluation part of the proposal.

3.1. Linear Regression and Decision Trees

The goal of linear regression models is to find a linear mapping between observed features and observed real outputs so that, when a new instance is seen, the output can be predicted [27]. Regression is a method of modeling a target value based on independent predictors. This method is mostly used for forecasting and finding out cause and effect relationships between variables. DTs are decision tools based on a tree-model [28,29]. They are navigated from the root to the leaves, each intermediate node represents a decision point and the ramification represents the properties that leads to a particular decision. The predicate that is associated with each internal node, which is used to discriminate among the data, is called "split condition". When a leaf is reached by navigating the tree, not only is a particular classification associated with the input instance, but, thanks to the path, it is possible to understand the reason for a particular result. A DT should be used when the relations among the various aspects of a specific application context are difficult to explain. In this case, the nonlinear approach of the DT performs better than the Linear Regression.

3.2. Support Vector Machines and Naive Bayes

SVMs are linear models for classification and regression problems which are used to solve linear and nonlinear, problems [30,31]. The idea of SVM is based on the definition of a line or a hyperplane which separates the data into classes. Based on given labeled input, the algorithm outputs a hyperplane-based model that is able to classify new instances. Given a set of training examples (training set), each of which is labeled with the class to which the two classes belong, an SVM training algorithm constructs a model that assigns new examples to one of the two classes, thus obtaining a non-probabilistic binary linear classifier. An SVM model is a representation of the examples as points in space, mapped in such a way that the examples belonging to the two different categories are clearly separated by the widest possible space. The new examples are then mapped in the same space and the prediction of the category to which they belong is made on the basis of the side in which it falls. In addition to linear classification, it is possible to use SVM to effectively perform nonlinear classification using the kernel method, implicitly mapping their inputs into a multi-dimensional feature space. NBs belong to a probabilistic family of classifiers [32]. They are centered on the theorem of Bayes, which is based on the assumptions of independence among features. NBs are highly scalable, requiring a number of parameters linear in the number of variables. Furthermore, the method is very efficient for text categorization, which can compete with more advanced methods including SVMs, with appropriate pre-processing.

3.3. Random Forest and Random Committee

The Random Forest is a very popular algorithm for feature ranking [33]. It belongs to the Bagging methods (Boostrap Aggregating) and it is based on the use of multiple decisions trees DT_k, each of which is trained on a subset S_k of the training set. Each new instance provided in input is classified by all the k-Decision Trees, each of which provides its own classification. A voting mechanism, which can consist of majority vote rule or on the average value gathered from all the k-classifications, is then adopted to establish the final classification based on the most common class in the node. However, the Random Committee builds an ensemble of base classifiers and generates their prediction by averaging of the estimated probability [34]. Each base classifier is based on the same data, but it uses a different random number seed, which makes sense if the base classifier is randomized; otherwise, all classifiers would work in exactly the same way.

3.4. Boosting and Bagging

Bagging and Boosting are also ensemble methods [35,36]. The idea of Boosting is to combine "weak" classifiers in order to create a classifier with a better accuracy. In algorithms such as Adaboost, the output of the meta-classifier is given by the weighted sum of the predictions of the individual models. Whenever a model is trained, there will be a phase of repeating the instances. The boosting algorithm tends to give greater weight to the misclassified instances, with the aim of obtaining an improved model on the basis of these latter instances. On the contrary, the Bagging approach aims to reduce variance from models that might have a very high level of accuracy, but typically only with the data, on which they have been trained, which is called over-fitting. It tries to reduce this phenomenon by creating its own variance among the data by sampling and replacing data by testing diverse models called hypothesis.

3.5. Artificial Neural Networks

ANNs are particular computational models, which are able to represent knowledge based on massive parallel processing and pattern recognition based on past experience or examples [37]. An ANN is defined through an initial layer on the basis of the available inputs, a final layer which represents the output of the computation and a hidden layer which is defined in terms of potential multi-layers through which the inputs undergo various transformation and calculation steps as long as the final layer is reached and the output is generated. They are computation models inspired by biological networks in which: (i) the information processing occurs at several simple elements that are called neurons; (ii) signals are passed between neurons over connection links; (iii) each connection link has an associated weight, which, in a typical neural net, multiplies the signal transmitted; (iv) each neuron applies an activation function (usually nonlinear) to its net input (sum of weighted input signals) to determine its output signal. By such replicated learning process and associative memory, an ANN model can classify information as pre-specified patterns. A typical ANN consists of a number of simple processing elements called neurons, nodes or units. Each neuron is connected to other neurons by means of directed communication links. Each connection has an associated weight, which represents the parameters of the model being used by the net to solve a problem. ANNs are usually modeled into one input layer, one or several hidden layers, and one output layer.

4. Research Approach Description

This section aims to clarify the approach adopted in this research task, which is depicted as a process in Figure 2. The process is designed in three main parts, which are organized in lanes: "Data Management", "Phase" and "Work-product". More specifically, the "Data Management" lane is related to the dataset elaboration, the "Phase" lane provides the information about what is done and in which order, whereas the "Work-product" lane illustrates which output is generated and how it is eventually used. By describing the process by phase from the top to the bottom, its functioning as well as the sequence order of its actions are highlighted. In particular:

- *Model definition*: this phase starts by taking in as input an *Initial Dataset* which contains a collection of data related to different types of devices. This phase aims at identifying common features among the different type of devices that will be used to characterize and discriminate them. The output of this phase is represented by a model based on different features called *Feature-based model*.
- *Feature-driven value extraction*: this phase uses both the *Initial Dataset* and the *Feature-based model*. In particular, the *Feature-based model* is applied to the *Initial Dataset* and, in particular, on the recorded traces in order to extract additional information, called *Derived Dataset*. Such information, which enriches the traces available in the *Initial Dataset*, is then exploited to distinguish the different appliances.
- *Data splitting*: this phase is centered on the *Derived Dataset* and aims to divide it into two disjoint subsets: *Training Set* and *Test Set*. The *Training Set* is provided in input to a learning algorithm in order to train appropriately the classifier which is built on the basis of the identified features, whereas the *Test Set* is used to validate it.
- *Learning*: in this phase, one or more learning algorithms are chosen on the basis of both the analysis to be conducted (e.g., supervised, semi-supervised or non-supervised) and the kind of available data (e.g., labeled or not labeled). This phase ends by training such learning algorithm by using the *Training Set* in order to obtain a *Trained classifier* from each of them.
- *Model validation*: in this phase, the capacity of the trained classifiers obtained in the previous phase in the identification and classification of electrical devices is assessed. To this aim, only the *Test Set*, which consists of traces of devices that have been not used to train the classifiers, are employed so as to get the *Classification Results*.
- *Feature analysis*: this phase ends the overall process. In particular, starting from the *Classification Results* gathered from the previous phase, the features that play the most important role, in the electrical devices identification, are discussed.

In the next sections, more in-depth details on the conducted research activity are given, by focusing on the work-products. Specifically, Section 5 provides the full description of the *Feature-based model* by describing a particular instance of it called EDIM (Electrical Devices Identification Model) by highlighting which aspects have been considered in its definition and why, whereas the *Trained classifiers*, the *Classification Results* and the *Importance Feature Results Analysis* are discussed in Section 6.

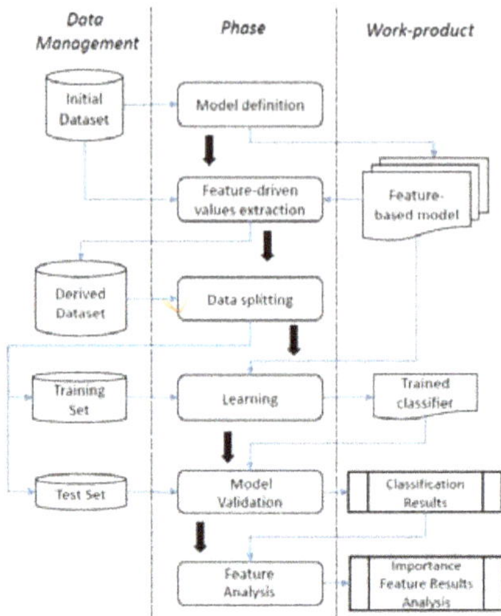

Figure 2. Research approach: data management, phases and work-products.

5. EDIM: An Electrical Devices Identification Model and Related Features

In this section, the proposed *Electrical Devices Identification Model (EDIM)* for the identification of electrical devices is described. Specific aspects, related to the usage of appliances in terms of time, place and electricity consumption have been considered for its definition. Their combination aims to highlight emergent behaviors that are able to characterize and discriminate among different devices.

In particular, as it is depicted in Figure 3, the *EDIM* model is inspired by three main driving questions: (i) HOW MUCH does a device consume? (ii) WHEN does a device consume? (iii) WHERE is a device used? The rationale behind them relies not only on extracting and using information that is directly derivable or measurable from a device, such as its energy consumption, but also to combining it with further information related to the way of using a particular device—for example, by considering differences of the use of a device in specific daily time slots, weekly or seasonal, relationships with other devices such as their use in sequence or in parallel, as well as by distinguishing whether a device is used in a specific area or if it is used, with a certain frequency, in different places.

By dealing with the above-mentioned questions, three main related *feature classes* have been identified, namely *Energy and Power Consumption, Temporal Usage* and *Appliance Location*. For each of them, a set of specific features have been proposed, for a total of 19 basic features. In the next subsections, the description of the defined *feature classes* is provided and, for each of them, the features that have been proposed are presented and formalized through a mathematical notation.

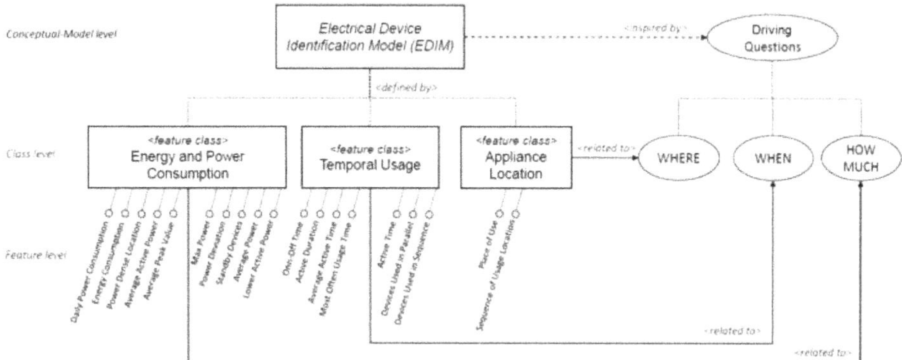

Figure 3. The Electrical Device Identification Model and related questions.

5.1. Energy and Power Consumption Class

This feature class focuses basically on the measurement of power and energy at various levels and at specific points of time. The features belonging to this class aim to extract information related to the electricity consumption of a device in order to characterize it, by answering the question "HOW MUCH does a device consume?" In this class, the following consumption-related features are identified: *Daily Power Consumption, Max Power, Power Deviation, Average Power, Average Active Power, Lower Activity Power (or MinPower), Energy Consumption, Average Peak Value, Power Dense Location,* and *Standby Devices.* In the next section, the above-mentioned features, which have been proposed in this class, are elaborated.

Daily Power Consumption. It is used to compute the amount of power p consumed by a device j on a day D, by observing it in a time unit $i[s] \in D$:

$$P_j^{Tot} = \sum_{i \in D} p_j(i), \text{where } i \text{ corresponds to a second [s]}. \tag{1}$$

Max Power. This feature is used to calculate the maximum power value p used by a device j within a specific day D:

$$P_j^{Max} = max\{p_j(i)\}, \text{where } i[s] \in D. \tag{2}$$

Power Deviation. This feature deals with the power deviation which is computed as the sum of the difference between the *Max Power* of a device j within a reference day D and its power consumption at every time unit $i[s] \in D$. Only when j is in operation $Status_j(i) > \lambda = 5[W] = On$:

$$P_j^{Dev} = \sum_{i \in D} (P_j^{Max} - p_j(i)) \Leftrightarrow Status_j(i) = On. \tag{3}$$

Average Power. Given a device j, this feature calculates the average power used from it, related to a day D by considering both active and non-active operation time $i \in D$:

$$P_j^{Avg} = \frac{\sum p_j(i)}{count(i)}, \forall i \in D. \tag{4}$$

Average Active Power. Given a device j, this feature calculates the average power used from it, related to a day D by considering only the active operation time $i \in D$, such that $Status_j(i) > \lambda = 5[W] = On$:

$$P_j^{AvgAct} = \frac{\sum p_j(i)}{count(i)}, \forall\, i \in D \text{ s.t. } Status_j(i) = On. \tag{5}$$

Lower Activity Power (Min Power). This feature is used to calculate the minimum power value p used by a device j within a specific day D, by considering only the active operation time $i \in D$, such that $Status_j(i) > \lambda = 5[W] = On$:

$$P_j^{Min} = min\{p_j(i)\}, \text{where } i[s] \in D \text{ s.t. } Status_j(i) = On. \tag{6}$$

Energy Consumption. Given a time period D (e.g., a Day, a Week, a Month) divided into a set of n sub-periods $\{d_1, d_2, ..., d_n\} \subset D$. This feature is used to calculate the energy consumption of a device j in a specific sub-period $b \in D$:

$$EC_j^b = \sum_i^D p_j(i) \Leftrightarrow i \in b \subset D. \tag{7}$$

Average Peak Value. Given a reference period of time D_j (e.g., a Day, a Week, a Month) as a disjoint list of $K = \{1, 2, ..., k\}$ time intervals $I_j = \{sp_j(h_0, h_1), sp_j(h_2, h_3), ..., sp_j(h_{k-1}, h_k)\}$ in which a device j was actively used, then the Average Peak Value of a device j APV_j calculates the average of all the peak values within the considered period of time D_j, where $peak(sp_j(h', h'')) = max\{p_j(i)\}$ with $h' < i < h''$ is the max value of energy consumed from the the device j in the time interval $[h', h'']$:

$$APV_j(D_j) = \frac{\sum peak(sp_j(h', h''))}{k}, \forall\, sp_j(h', h'') \in I_j. \tag{8}$$

Power Dense Location. Given a location l and a set of n devices $J = \{j_1, j_2, ..., j_n\}$. This feature provides the amount of power consumed in l from all the devices in J in an arbitrary period of time D, if the total power consumed is more than a reference threshold $PD_{threshold}$:

$$PD_D^l = \sum_j^J \sum_{i=0}^T p_j^l(i) > PD_{threshold}. \tag{9}$$

Standby Devices. Given a set of n devices $J = \{j_1, j_2, ..., j_n\}$, this feature calculates a subset of devices $SB_{Dev} = \{j_1, j_2, ..., j_k\} \subset J$ which are neither Off nor On, rather those which present a standby mode that is a power consumption $0 < p_j(i) < \lambda = 5[W]$ for at least an uninterrupted period of time δt:

$$SB_{Dev} = < j_1, j_2, ..., j_k >, \text{if } \exists\, p_j(i) \text{ s.t. } 0 < p_j(i) < \lambda, \text{ and } continuous(i) > \delta t. \tag{10}$$

5.2. Temporal Usage Class

This feature class focuses on the use of a device, mainly from a temporal point of view, by considering the question "WHEN does a device consume?" The features that fall into this class try to extract information, regardless of the amount of energy consumed, with the aim of identifying temporal usage patterns such as daily, weekly, seasonal related to a single device as well as sequence-parallel relationships between multiple devices (e.g., the dryer after the washing machine, or the decoder along with the television). In this class, the following time-related features are identified: *On-Off Time, Active Time,*

Average Active Time, Active Duration, Most often Usage Time, Devices Used in Sequence, Devices Used in Parallel. The above-mentioned features, which have been proposed in this class, are described below in more detail.

On-Off Time. This feature is used to know, in which instant of time i of a day D, a device j is turned On/Off. The function $Status_j(i)$ is used to check the working status of j at the time i [s] $\in D$, in order to identify when a change occurs, based on the previous instant of time $i - 1$, where λ=5[W] is the On-threshold:

$$T_j^{On-Off}(i) = \begin{cases} Off, & when\ \text{Status}_j(i-1) > \lambda\ and\ \text{Status}_j(i) = 0, \\ On, & when\ \text{Status}_j(i-1) = 0\ and\ \text{Status}_j(i) > \lambda. \end{cases} \tag{11}$$

Active Time. This feature counts the number of times that a device j is turned on in an arbitrary day D. For example, a dishwasher is typically turned on once or twice a day:

$$T_j^{Act}(D) = count(T_j^{On-Off}(i)) \Leftrightarrow T_j^{On-Off}(i) = On,\ \forall\ i \in D. \tag{12}$$

Active Duration. Given a device j, its active duration in an arbitrary day D represents the overall time i in which j is active that is $Status_j(i) = On$:

$$T_j^{ActiveDur}(D) = count_j(i) \Leftrightarrow Status_j(i) > \lambda = 5[W] = On,\ with\ i \in D. \tag{13}$$

Average Active Time. This feature calculates the active average duration of a device j within an arbitrary day D:

$$T_j^{AvgAct}(D) = \frac{T_j^{ActiveDur}(D)}{T_j^{Act}(D)}. \tag{14}$$

Most Often Usage Time. Given a reference period of time D_j (e.g., a Day, a Week, a Month) as a disjoint list of $K = \{1, 2, ..., k\}$ time intervals $I_j = \{sp_j(h_0, h_1), sp_j(h_2, h_3), ..., sp_j(h_{k-1}, h_k)\}$ in which a device j was actively used, then the most often usage time of a device j indicates the longest interval of time $\Delta h_j = < h', h'' >_j$ such that $sp_j(h', h'') \in I_j$ and $h', h'' \in K$, in which the device j was used:

$$\Delta h_j = max < h', h'' >_j = max\{(h'' - h')_j\} = max\{sp_j(h', h'')\}. \tag{15}$$

Devices Used in Sequence. Given two instants of time i_1 and i_2 with $i_2 \geq i_1$. A device j_2 works in sequence after j_1 $seq((j_2, j_1))$, when j_1 stops working at time i_1, which is $Status_{j1}(i_1 - 1) = On$ and $Status_{j1}(i_1) = Off$ and the device j_2 starts working in a subsequent instant of time i_2, which is $Status_{j2}(i_2) = Off$ and $Status_{j2}(i_2 + 1) = On$. Thus, this feature $S\bar{E}Q_J(D)$ returns all the possible couples of devices $j_1, j_2 \in J$, which work in sequence in a reference period D:

$$S\bar{E}Q_J(D) = \{seq(j_1, j_2)\}_D\ \forall\ j_1,\ j_2 \in J\ set\ of\ devices. \tag{16}$$

Devices Used in Parallel. Given two instants of time i_1 and i_2 with $i_1 > i_2$, $i_1 > (i_2 + 1) - \Delta t$ and $\Delta t \geq 1$. The device j_2 works in parallel with j_1 $par((j_2, j_1))$, when j_1 stops working at time i_1 that is $Status_{j1}(i_1 - 1) = On$ and $Status_{j1}(i_1) = Off$ and the device j_2 starts working in an instant of time i_2 that is $Status_{j2}(i_2) = Off$ and $Status_{j2}(i_2 + 1) = On$. This feature $P\bar{A}R_J(D)$ returns all the possible couples of devices $j_1, j_2 \in J$, which work in parallel at least for a threshold Δt in a reference period D:

$$P\bar{A}R_J(D) = \{par(j_1, j_2)\}_D\ \forall\ j_1,\ j_2 \in J\ set\ of\ devices. \tag{17}$$

5.3. Appliance Location Class

The place of use of a device is another important indicator, since some electrical devices are often used in the same place (e.g., the hairdryer in the bathroom, the kettle in the kitchen). Some of them are movable and others are not. As a consequence, some devices can be used in more than one location inside a house and they can be active in not more than one location at a time. This feature class is driven by the question "WHERE is a device used?" Considering this aspect, in this class, the following location-related features are identified: *Place of Use, Sequence of Usage Location.*

Sequence of Usage Location. Given a set of locations $L = \{l_1, ..., l_z, ..., l_k\}$ that represent specific places (e.g., a kitchen, a bathroom, a bedroom, a living room and so on) in a given house h. The feature computes the list of locations in a house h, where a device j was chronologically used $p_j^{l_z}(i_w) > \lambda = 5[W] = On$ in an arbitrary day D:

$$SoUL_j^h(D) = < l_1(i_1), ..., l_z(i_w), l_{z+1}(i_{w+1}), ..., l_k(i_w) >_j^h, \text{with } i_1...i_w \in D, \tag{18}$$

$\forall <i_w, i_{w+1}>$ with $i_w < i_{w+1}$ and $p_j^{l_z}(i_w) > \lambda$ and $p_j^{l_{z+1}}(i_{w+1}) > \lambda$.

Place of Use Given a set of locations $L = \{l_1, l_2, ..., l_k\}$ that represent specific places (e.g., a kitchen, a bathroom, a bedroom, a living room) in a given house h. This feature allows for knowing in which location $z \in L$ the device j was used $p_j^z(i) > \lambda = 5[W] = On$. That is, it was On for at least one time unit i in an arbitrary day D:

$$z_j^h(i) \Leftarrow p_j^z(i) > \lambda, \text{ with } z \in L \text{ and } i \in D. \tag{19}$$

6. Experiments and Results Discussion

In this section, first an overview of the used dataset is given, then the results gathered by experimenting the proposed model are described and, finally, a discussion on the importance of the features is provided.

6.1. Dataset Overview

The dataset used to evaluate the proposal consists of a collection of traces related to the daily use of different electrical devices. This dataset, which is made available under the Open Database License (ODbL) [38], is public available and freely downloadable [14]. Each entry of the dataset contains basic data such as the identifier of a device, the time unit with a granularity of a second, which is used to collect the data of each device, the amount of energy consumed in a time unit and so on.

The trace-base data have been grouped into three categories in accordance with the process they have been collected: "Full-day traces": which contains complete traces that have been recorded for a time period over 24 h; "Incomplete traces": which contains traces with missing measurements in different instants of time over the day; "Synthetic traces": which contains trace fragments of devices that have been manually completed with values corresponding to zero consumption readings, when the real values were not available. In our case, the folder containing the "Full-day traces" has been used in the experimentation, which corresponds to 33 types of devices, as listed in Table 3.

6.2. Features Evaluation

A machine learning based approach has been used for the evaluation of the proposed identification model. Starting from such row data available in the "Full-day traces" folder, additional information has been extracted by using the features that have been proposed in Section 5. Both the basic and extracted information is used to train and test different classifiers. As the dataset is labeled, the case in consideration falls under a supervised learning problem. As a consequence, only supervised machine learning techniques have been considered and compared. In particular, Random Forest, Bagging, LogitBoost, Decision Tree, Naive Bayes and SVM algorithms have been selected and experimented with for two main reasons: (i) on one hand, to the best of our knowledge, they have shown the best performance in literature, (ii) on the other hand, it is possible to analyze and understand the logic behind their classification process, which is, instead, not always possible with other techniques. For example, neural networks make difficult to understand what happens during the classification, since they act as a black box, to understand which features play the most important role for the device identification. As a consequence, they have not been considered. For the sake of completeness and clarity of this work, Figure 4 summarizes the machine learning algorithms that have been used in the experiments along with the values of the parameters that have been used after having tuned them.

Bagging	
Parameter	*Value*
n_estimators	10
max_samples	1
max_features	1
bootstrap	WAHR
bootstrap_features	FALSCH
oob_score	FALSCH
warm_start	FALSCH
n_jobs	none
random_state	none
verbose	0

Random Forest	
Parameter	*Value*
n_estimators	100
criterion	gini
max_depth	None
min_samples_split	2
min_samples_leaf	1
min_weight_fraction_leaf	0
max_features	Auto
max_leaf_nodes	None
min_impurity_decrease	0
min_impurity_split	1.00E-07
bootstrap	WAHR
oob_score	FALSCH
n_jobs	None
random_state	None
verbose	0
warm_start	FALSCH
class_weight	None

Decision Tree	
Parameter	*Value*
criterion	gini
splitter	best
max_depth	none
min_samples_split	2
min_samples_leaf	1
min_weight_fraction_leaf	0
max_features/random_state	none
max_leaf_nodes	none
min_impurity_decrease	0
min_impurity_split	1.00E-07
class_weight	none
persort	FALSCH

Logit Boost	
Parameter	*Value*
n_estimators	100
weight_trim_quantile	0.05
max_response	4
learning_rate	1
bootstrap	FALSCH
random_state	none

Naïve Bayes	
Parameter	*Value*
Priors	None
v_smoothing	1.00E-09

Support Vector Machine	
Parameter	*Value*
gamma	scale
kernel	rbf
shrinking	true

Figure 4. Machine learning algorithms and related parameters.

The training and test set have been constructed by using a standard approach based on 5-fold cross validation, so as to reduce both the risk of losing important patterns/trends in data set and the error induced by bias. In more detail, 80% of all 33 categories of available devices' traces have been exploited to build the training set extracted from the traces, with the remaining 20% of the data to build the test set, which are employed respectively to train and test the above-mentioned classifiers. Moreover, the testing of the model was done for each single device in order to get the prediction of each device separately (i.e., at device level), and then the accuracy of all the devices were averaged to calculate the overall accuracy of each trained model. Moreover, as different traces of different models of certain device categories were available (for example LCD-TV, Router, Washing Machine), the construction of the training set and the test set took into account that the traces belonging to a certain device model were used only in the training set or in the test set. This allowed for showing that the proposed features are able to recognize even new devices with similar behavior, in terms of energy usage, belonging to one of the 33 device categories under consideration.

A first result is shown in Table 2 in terms of accuracy of the different classifiers. A general observation is that all the classifiers present an accuracy higher than 90%, which provides an indication of the goodness of the selected features. Indeed, they show a certain degree of independence from a specific classifier and, as a consequence, they are well suited to describe appliance types and distinguish them from others. Moreover, among them the best performance is reached by the Random Forest algorithm with 96.51% accuracy, which is why the rest of the analysis is specifically based on its use for the subsequent evaluation. The details are reported in Table 3 by showing the result values for *true* and *false positives* as well as the *precision*, which measures the proportion of actual negatives that are correctly identified as such, and the *recall*, which measures the proportion of actual positives that are correctly identified as such, for each kind of device.

Table 2. Accuracy of the different classifiers.

#	Algorithm	Accuracy [%]
1	Random Forest	96.51
2	LogitBoost	94.99
3	Bagging	93.02
4	Decision Tree	91.10
5	Naive Bayes	90.26
6	Support Vector Machine	90.11

As we can see, our implementation reaches at least 80% accuracy and almost all the devices are always classified correctly, which can be seen from the true positive ratio, which is equal to 1.0. However, for other devices, the false positive ratio is, however, very low. Additionally, both the results obtained by calculating the precision and the recall values comply with the observed values related to the true positive rate. Only for one device, and in particular for the Water Kettle, a lower level of classification is shown. This is associated with the limited number of instances available during the training phase of the classifier that made the training phase of the classifier very difficult for this typology of device.

In general, some devices present electrical characteristics that are easier to recognize and which require few instances for training the classifier; others instead require a greater number of instances. This can be traced back to the fact that the behavior of some devices is not only strongly dependent on their mode and state of operation, such as for an alarm clock or a vacuum cleaner. Others, instead, have dependencies on their state and external factors, for example in the case of the Water Kettle, the amount of water to be heated could influence the duration of its heating process. Consequently, not only is it necessary to have a sufficient amount of traces, but they should also be collected considering these additional factors.

However, more than 60% of the devices are correctly identified and classified, as we can see from the true positive rate, precision and recall, without errors. In summary, we globally obtained very high performances in terms of accuracy compared with other related works, and in particular with respect to the related work [15], by only using 19 features instead of 517 features on the same dataset, by requiring both less computational resources and computing time.

Table 3. Accuracy classification for each appliance provided by the Random Forest.

Electrical Device	True Positive	False Positive	Precision	Recall
Alarm Clock	1.0	0	1.0	1.0
Amplifier	1.0	0	1.0	1.0
Bean to cup	1.0	0	1.0	1.0
Coffee machine	1.0	0	1.0	1.0
Dishwasher	1.0	0	1.0	1.0
Desktop PC	1.0	0	1.0	1.0
Dryer	1.0	0	1.0	1.0
DVD	0.99	0.001	0.941	0.99
Ethernet	0.95	0	1.0	0.95
Freezer	1.0	0	1.0	1.0
Iron	0.80	0.002	0.65	0.80
Lamp	0.88	0.002	0.85	0.88
Laptop	0.96	0	1.0	0.96
Mediacentre	0.99	0	1.0	0.99
Microwave	1.0	0.001	0.95	1.0
Monitor-CRT	0.92	0.002	0.93	0.92
Monitor-TFT	1.0	0	1.0	1.0
PlayStation	0.87	0	1.0	0.87
Printer	1.0	0.001	0.98	1.0
Projector	0.97	0	1.0	0.97
Refrigrator	1.0	0	1.0	1.0
Router	1.0	0	1.0	1.0
Stove	1.0	0	1.0	1.0
Toaster	1.0	0.001	0.95	1.0
TV-CRT	1.0	0	1.0	1.0
TV-LCD	1.0	0	1.0	1.0
TV-REC	0.96	0	1.0	0.96
USB Harddrive	1.0	0	1.0	1.0
Vacuum cleaner	1.0	0	1.0	1.0
Water Fountain	1.0	0	1.0	1.0
Water Kettle	0.57	0.003	0.58	0.57
Wash Machine	1.0	0.002	0.983	1.0
Xmas Lights	0.99	0	1	0.99
Weighted average	0.9651	0.0004	0.964	0.9651

6.3. Discussion of the Features Importance

In addition, not only is it important to have a model that performs well, but it is also very important to understand why it works good (or bad) and under which conditions. This helps to understand the logic of the model and the reasoning behind a decision. Knowing the importance of a feature in the classification process may motivate the exploitation of more complex one or removing them based on their significance, even by sacrificing some accuracy for the sake of the interpretability. In our case, an analysis on the feature importance has been conducted, in order to know which of them plays the most important role in the electrical device identification process.

In the assessment of the feature importance, a common evaluation criterion is called impurity, which is used to express the level of homogeneity (or heterogeneity) among a group of items [39]. In our case, the classification model is based on the Random Forest, which in turn consists of a set of sub-trees. In this case, the impurity value for each feature is calculated on the basis of each sub-tree. As a consequence, the impurity is assessed over all the nodes for all the trees. Every node in the decision tree splits the dataset into two subsets, so that all the results showing similarities fall in the same subset. As a consequence, from one side, the more important a feature is, the more it decreases the impurity in the tree, but, on the other side a mistake based on it will produce a greater impact on the overall classification. Typically, the features which generate greatest decrease in terms of impurity are located closer to the root-level of a tree, whereas those that produce less decrease of impurity are closer to the leaf-level.

In this case, Table 4 shows how much each feature contributes in the reduction of the level of impurity and, as a consequence, their importance in percentage in the identification of the electrical devices. As it is visible, the features are sorted by level of importance. The most important feature is shown in the first row of the table. Moving down, the less important features are listed. In particular, the *Avarage Peak Power* is the most important one because it provides the highest level of impurity reduction, unlike with the *Energy Consumption*, which is the feature that contributes the least to discriminating between appliances and, consequently, to reduce the level of impurity among them. It is important to note that no value of importance is reported regarding to the *On-Off Time*. This is because it does not contribute directly as a discriminating characteristic, but it is used indirectly from other more complex features in the classification process, as a supporting feature.

Table 4. Descending order of features, for level of importance, expressed as a percentage.

#	Feature Name	Importance (I) [%]
1	Average Peak Power	14.8142
2	Average Active Power	14.6424
3	Max Power	11.3404
4	Average Active Time	10.1315
5	Lowest Active Power	10.1219
6	Place of Use	8.366
7	Active Duration	7.6177
8	Devices Used in Parallel	5.2789
9	Average Power	4.9267
10	Most of Usage Time	4.5993
11	Standby Devices	2.0524
12	Active Time	1.9884
13	Power Deviation	1.7767
14	Devices Used In Sequence	1.2207
15	Power Dense Location	0.6378
16	Sequence of Usage Location	0.2529
17	Daily Power Consumption	0.2319
18	Energy Consumption	0.0002
19	On-Off Time	-
Tot.	-	100

A further observation can be made by calculating the average percentage value of importance of a feature $\bar{I}_F = 5.26\%$ as:

$$\bar{I}_F = \frac{100}{count(F)} = \frac{100}{19} = 5.26\%, \quad \text{with F the set of all features.} \tag{20}$$

On the basis of such reference parameter $\bar{I}_F$, only the first eight features of Table 4 present a higher value than $\bar{I}_F$. It means that *Average Peak Power, Average Active Power, Max Power, Average Active Time, Lowest Active Power, Place of Use, Active Duration, Devices Used in Parallel* can be considered the most important features in the classification process. Indeed, by summing their percentage values of importance, the result is equal to 82.31%, which reflects the level of accuracy of at least 80%, in the classification of almost all the devices with a true positive ratio equal to 1.0, as described above. However, the diagram depicted in Figure 5 reports for such features both the absolute importance values with respect to all 19 features as well as the relative values when only the most important ones are considered.

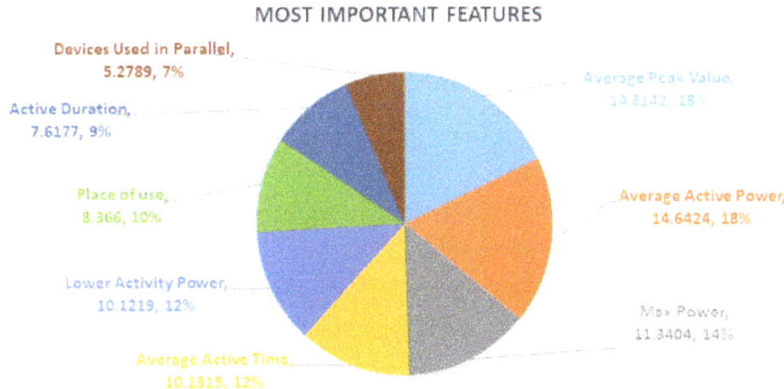

Figure 5. The most important features along with their absolute and relative importance values.

7. Conclusions

The paper focused on the automatic identification of electrical devices based on features. An *Electrical Device Identification Model* centered on three feature classes related to energy consumption, time usage and location, have been proposed. For each class, specific features have been defined and formalized for a total of 19 distinct features. The information extracted by applying such features has been used to train six different classifiers (i.e., Random Forest, Bagging, LogitBoost, Decision Tree, Naive Bayes and SVM), which have shown a high level of accuracy as a symptom of goodness of the proposed features. Of course, a number of variations of techniques, which are designed for different types of tasks, are also available. From one side, they typically allow for obtaining even better results; this means that, to some extent, they influence the assessment. Since our aim was to evaluate the proposed features by avoiding any potential kind of bias, in this research activity, we considered only standard techniques in order to obtain a more neutral assessment of the features and to compare the results with the prior works reported in Section 2. Further experiments and details related to the Random Forest classifier that provided the highest accuracy equal to 96.51% have been conducted and discussed, as it outperformed with respect to the related work [15] by using only 19 features on the same dataset. In particular, the ratios of true and false positives, as well as the precision and the recall with reference to the specific appliances, have been evaluated. An additional analysis has been done, in order to understand the logic of the classifier and the reasoning behind its decisions. Specifically, not only did it emerged how much each feature contributes in the classification process, but also the most important ones have been identified.

Ongoing works aim to extend this identification model in order to (i) enhance the interaction between Smart Homes and Smart Grids to improve the decision-making process for the energy automatic management and distribution in the network, and (ii) improve the local management of electrical devices in smart homes automatically on the basis of users' habits, and centered on the definition of specific user' profiles. Furthermore, it is worth noting that the behavior of some devices might change over time due to aging, temperature and environmental effects. For example, with aging, the battery of a smart phone might show performance degradation in terms of needed energy while charging or it might take more time to charge or might discharge at a faster rate. Similarly, it could happen with other devices like an air conditioner whose usage changes from season to season, also depending on the external temperature. Such behaviors cannot be covered in a limited time trace dataset, which is why there could be some points where the classification fails. As the devices' behavior or their use can change over the time, a possible future work regards the extension of our model with a patching-based approach for classifiers [40], which focuses on the adaption of existing classification models to new data. As classification often faces scenarios where an already existing model needs to be adapted to a changing environment. Such research work identifies the regions of instance space in which adaptation is needed and, after that, local classifiers for these regions are trained. Such regions after training are incorporated into the model and can handle the predictions even after some changes due to aging or environment. Thus, whenever a decay is detected in the performance of the model, the adaptation is triggered with the goal of finding patches to the classifier that can act efficiently without training the model again from scratch.

Author Contributions: The authors contributed equally in all parts of the article in terms of literature review, adopted methodology, feature identification, model definition, experimentation and results analysis.

Conflicts of Interest: The authors declare no conflict of interest.

Abbreviations

The following abbreviations are used in this manuscript:

SG	Smart Grid
SGs	Smart Grids
SH	Smart Home
SHs	Smart Homes
EDIM	Electrical Devices Identification Model
MAUs	Measurement and Actuation Units
ODbL	Open Database License
ML	Machine Learning
DT	Decision Tree
LR	Linear Regression
SVM	Support Vector Machine
NB	Naive Bayes
RF	Random Forest
RC	Random Committee
ANN	Artificial Neural Network
NIALM	Non Intrusive Appliance Load Monitoring

References

1. Karnouskos, S. Cyber-Physical Systems in the SmartGrid. In Proceedings of the 2011 9th IEEE International Conference on Industrial Informatics, Lisbon, Portugal, 26–29 July 2011; pp. 20–23.
2. IEEE. *IEEE Vision for Smart Grid Control: 2030 and Beyond Reference Model*; IEEE: Piscataway, NJ, USA, 2013; pp. 1–10. [CrossRef]
3. Eurostat. 2018. Available online: https://www.statista.com/statistics/418078/electricity-prices-for-households-in-germany (accessed on 28 February 2019).
4. Egert, R.; Tundis, A.; Roth, S.; Mühlhäuser, M. *A Service Quality Indicator for Apriori Assessment and Comparison of Cellular Energy Grids*; Kaparaju, P., Howlett, R.J., Littlewood, J., Ekanyake, C., Vlacic, L., Eds.; Sustainability in Energy and Buildings 2018; Springer International Publishing: Cham, Switzerland, 2019; pp. 322–332.
5. Rogovchenko-Buffoni, L.; Tundis, A.; Hossain, M.Z.; Nyberg, M.; Fritzson, P. An integrated toolchain for model based functional safety analysis. *J. Comput. Sci.* **2014**, *5*, 408–414. [CrossRef]
6. Tundis, A.; Egert, R.; Mühlhäuser, M. Applying a Properties Modeling approach for monitoring Smart Grids. In Proceedings of the 2017 IEEE 14th International Conference on Networking, Sensing and Control (ICNSC), Calabria, Italy, 16–18 May 2017; pp. 714–719. [CrossRef]
7. Mahmood, A.; Butt, A.R.; Mussadiq, U.; Nawaz, R.; Zafar, R.; Razzaq, S. Energy sharing and management for prosumers in smart grid with integration of storage system. In Proceedings of the 2017 5th International Istanbul Smart Grid and Cities Congress and Fair (ICSG), Istanbul, Turkey, 21 April 2017; pp. 153–156. [CrossRef]

8. Zafar, R.; Mahmood, A.; Razzaq, S.; Ali, W.; Naeem, U.; Shehzad, K. Prosumer based energy management and sharing in smart grid. *Renew. Sustain. Energy Rev.* **2018**, *82*, 1675–1684. [CrossRef]

9. Egert, R.; Cordero, C.G.; Tundis, A.; Mühlhäuser, M. HOLEG: A simulator for evaluating resilient energy networks based on the Holon analogy. In Proceedings of the 2017 IEEE/ACM 21st International Symposium on Distributed Simulation and Real Time Applications (DS-RT), Rome, Italy, 18–20 October 2017; pp. 1–8. [CrossRef]

10. Egert, R.; Tundis, A.; Volk, F.; Mühlhäuser, M. An integrated tool for supporting the design and virtual evaluation of smart grids. In Proceedings of the 2017 IEEE International Conference on Smart Grid Communications (SmartGridComm), Bejing, China, 21–24 October 2017; pp. 259–264. [CrossRef]

11. Alam, M.R.; Reaz, M.B.I.; Ali, M.A.M. A Review of Smart Homes—Past, Present, and Future. *IEEE Trans. Syst. Man Cybern. C (Appl. Rev.)* **2012**, *42*, 1190–1203. [CrossRef]

12. Smart Home: The Human Side of the Smart Grid. 2010. Available online: http://www.smartgrids-cre.fr/media/documents/1003-CapG-SmartHome.pdf (accessed on 28 February 2019).

13. Abeykoon, V.; Kankanamdurage, N.; Senevirathna, A.; Ranaweera, P.; Udawalpola, R. Real Time Identification of Electrical Devices through Power Consumption Pattern Detection. *Pervasive Comput.* **2016**, *10*, 40–48.

14. Tracebase. 2017. Available online: http://www.tracebase.org (accessed on 28 February 2019).

15. Reinhardt, A.; Baumann, P.; Burgstahler, D.; Hollick, M.; Chonov, H.; Werner, M.; Steinmetz, R. On the accuracy of appliance identification based on distributed load metering data. In Proceedings of the Sustainable Internet and ICT for Sustainability, Pisa, Italy, 4–5 October 2012; pp. 1–9.

16. Zufferey, D.; Gisler, C.; Khaled, O.; Hennebert, J. Machine learning approaches for electric appliance classification. In Proceedings of the 2012 11th International Conference on Information Science, Signal Processing and their Applications, Montreal, QC, Canada, 2–5 July 2012; pp. 740–745.

17. PLOGG—Energy Optimizers Limited—Plogg Wireless Energy Management. 2019. Available online: https://www.automatedhome.co.uk/new-products/plogg-appliance-energy-metering-and-control.html (accessed on 28 February 2019).

18. Morsali, H.; Shekarabi, S.M.; Ardekani, K.; Khayami, H.; Fereidunian, A.; Ghassemian, M.; Lesani, H. Smart plugs for building energy management systems. In Proceedings of the Iranian Conference on Smart Grids, Tehran, Iran, 24–25 May 2012; pp. 1–5.

19. Hart, G.W. Residential Energy Monitoring and Computerized Surveillance via Utility Power Flows. *Technol. Soc. Mag.* **1989**, *8*, 12–16. [CrossRef]

20. Ruzzelli, A.G.; Nicolas, C.; Schoofs, A.; O'Hare, G.M.P. Real-Time Recognition and Profiling of Appliances through a Single Electricity Sensor. In Proceedings of the 7th Annual IEEE Communications Society Conference on Sensor, Mesh and Ad Hoc Communications and Networks (SECON), Boston, MA, USA, 21–25 June 2010; pp. 1–9.

21. Patel, S.N.; Robertson, T.; Kientz, J.A.; Reynolds, M.S.; Abowd, G.D. At the Flick of a Switch: Detecting and Classifying Unique Electrical Events on the Residential Power Line (Nominated for the Best Paper Award). In Proceedings of the UbiComp, Innsbruck, Austria, 16–19 September 2007.

22. Ridi, A.; Gisler, C.; Hennebert, J. Automatic identification of electrical appliances using smart plugs. In Proceedings of the 8th International Workshop on Systems, Signal Processing and their Applications (WoSSPA), Algiers, Algeria, 12–15 May 2013; pp. 301–305.

23. Gulati, M.; Ram, S.S.; Singh, A. An in Depth Study into Using EMI Signatures for Appliance Identification. In Proceedings of the 1st ACM Conference on Embedded Systems for Energy-Efficient Buildings, Memphis, TN, USA, 4–6 November 2014; ACM: New York, NY, USA, 2014; pp. 70–79. [CrossRef]

24. HFED—High Frequency Energy Dataset. 2014. Available online: http://hfed.github.io/ (accessed on 6 May 2019).

25. Rashid, H.; Batra, N.; Singh, P. Rimor: Towards Identifying Anomalous Appliances in Buildings. In Proceedings of the 5th Conference on Systems for Built Environments, Shenzen, China, 7–8 November 2018; ACM: New York, NY, USA, 2018; pp. 33–42. [CrossRef]

26. Géron, A. *Hands-On Machine Learning with Scikit-Learn and TensorFlow: Concepts, Tools, and Techniques for Building Intelligent Systems*, 1st ed.; O' Reilly: Springfield, MO, USA, 2017.

27. Kavitha, S.; Varuna, S.; Ramya, R. A comparative analysis on linear regression and support vector regression. In Proceedings of the 2016 Online International Conference on Green Engineering and Technologies (IC-GET), Kuala Lumpur, Malaysia, 25–27 July 2016; pp. 1–5. [CrossRef]
28. Hu, Q.; Che, X.; Zhang, L.; Zhang, D.; Guo, M.; Yu, D. Rank Entropy-Based Decision Trees for Monotonic Classification. *IEEE Trans. Knowl. Data Eng.* **2012**, *24*, 2052–2064. [CrossRef]
29. Perner, P. How to Compare and Interpret Two Learnt Decision Trees from the Same Domain? In Proceedings of the 2013 27th International Conference on Advanced Information Networking and Applications Workshops, Barcelona, Spain, 25–28 March 2013; pp. 318–322. [CrossRef]
30. Han, B.; Davis, L.S. Density-Based Multifeature Background Subtraction with Support Vector Machine. *IEEE Trans. Pattern Anal. Mach. Intell.* **2012**, *34*, 1017–1023. [CrossRef] [PubMed]
31. Cheng, G.; Tong, X. Fuzzy Clustering Multiple Kernel Support Vector Machine. In Proceedings of the 2018 International Conference on Wavelet Analysis and Pattern Recognition (ICWAPR), Chengdu, China, 15–18 July 2018; pp. 7–12. [CrossRef]
32. Jahromi, A.H.; Taheri, M. A non-parametric mixture of Gaussian naive Bayes classifiers based on local independent features. In Proceedings of the 2017 Artificial Intelligence and Signal Processing Conference (AISP), Shiraz, Iran, 25–27 October 2017; pp. 209–212. [CrossRef]
33. Valecha, H.; Varma, A.; Khare, I.; Sachdeva, A.; Goyal, M. Prediction of Consumer Behaviour using Random Forest Algorithm. In Proceedings of the 2018 5th IEEE Uttar Pradesh Section International Conference on Electrical, Electronics and Computer Engineering (UPCON), Okinawa, Japan, 2–4 November 2018; pp. 1–6. [CrossRef]
34. Niranjan, A.; Prakash, A.; Veena, N.; Geetha, M.; Deepa Shenoy, P.; Venugopal, K.R. EBJRV: An Ensemble of Bagging, J48 and Random Committee by Voting for Efficient Classification of Intrusions. In Proceedings of the 2017 IEEE International WIE Conference on Electrical and Computer Engineering (WIECON-ECE), Uttarkhand, India, 18–19 December 2017; pp. 51–54. [CrossRef]
35. Khoshgoftaar, T.M.; Van Hulse, J.; Napolitano, A. Comparing Boosting and Bagging Techniques with Noisy and Imbalanced Data. *IEEE Trans. Syst. Man Cybern. A Syst. Hum.* **2011**, *41*, 552–568. [CrossRef]
36. Wang, B.; Pineau, J. Online Bagging and Boosting for Imbalanced Data Streams. *IEEE Trans. Knowl. Data Eng.* **2016**, *28*, 3353–3366. [CrossRef]
37. Kapoor, S.; Singh, A.; Goswami, U.; Kumar, S.; Chitranshi, G. Design and implementation of a robust system for recognizing alphabets using artifical neural network. In Proceedings of the 2016 5th International Conference on Reliability, Infocom Technologies and Optimization (Trends and Future Directions) (ICRITO), Noida, India, 7–9 September 2016; pp. 606–609. [CrossRef]
38. Open Database License. 2017. Available online: http://opendatacommons.org/licenses/odbl/1.0/ (accessed on 28 February 2019).
39. Xu, H.; Yang, M.; Liang, L. An improved random decision trees algorithm with application to land cover classification. In Proceedings of the 2010 18th International Conference on Geoinformatics, Beijing, China, 18–20 June 2010; pp. 1–4. [CrossRef]
40. Kauschke, S.; Fürnkranz, J. Batchwise Patching of Classifiers. In Proceedings of the 32nd AAAI Conference on Artificial Intelligence (AAAI-18), New Orleans, LA, USA, 2–7 February 2018; pp. 3374–3381.

Article

Development and Application of an Atmospheric Pollutant Monitoring System Based on LoRa—Part I: Design and Reliability Tests

Yushuang Ma [1], Long Zhao [1], Rongjin Yang [2], Xiuhong Li [1,3,*], Qiao Song [1], Zhenwei Song [1] and Yi Zhang [1]

[1] College of Global Change and Earth System Science, Beijing Normal University, No.19, Xinjiekou Wai Street, Haidian District, Beijing 100875, China; mays@mail.bnu.edu.cn (Y.M.); zhaolong_ak@163.com (L.Z.); zgsongqiao@163.com (Q.S.); szwzsx@126.com (Z.S.); 201721490029@mail.bnu.edu.cn (Y.Z.)

[2] Chinese Research Academy of Environmental Sciences, No.8, Da Yang Fang, An Wai, Chao Yang, Beijing 100012, China; yangrj@craes.org.cn

[3] State Key Laboratory of Remote Sensing Science, Jointly Sponsored by Beijing Normal University and Institute of Remote Sensing and Digital Earth of Chinese Academy of Sciences Beijing Normal University, Beijing 100101, China

* Correspondence: lixh@bnu.edu.cn; Tel.: +86-136-2116-6693

Received: 26 September 2018; Accepted: 8 November 2018; Published: 12 November 2018

Abstract: At present, as growing importance continues to be attached to atmospheric environmental problems, the demand for real-time monitoring of these problems is constantly increasing. This article describes the development and application of an embedded system for monitoring of atmospheric pollutant concentrations based on LoRa (Long Range) wireless communication technology, which is widely used in the Internet of Things (IoT). The proposed system is realized using a combination of software and hardware and is designed using the concept of modularization. Separation of each function into independent modules allows the system to be developed more quickly and to be applied more stably. In addition, by combining the requirements of the remote atmospheric pollutant concentration monitoring platform with the specific requirements for the intended application environment, the system demonstrates its significance for practical applications. In addition, the actual application data also verifies the sound application prospects of the proposed system.

Keywords: atmospheric; on-line monitoring; LoRa; embedded system

---

1. Introduction

Atmospheric pollution is an important issue in China because the severe effects of this pollution are detrimental to the life and health of the people [1]. The main sources of atmospheric pollution include organic compounds, carbon compounds, and particulate matter (PM) produced by human activities, including nitrogen dioxide, carbon monoxide and PM2.5, which can all seriously affect human health [2]. To provide better solutions to the problem of atmospheric pollution, it is essential to identify the source of each pollution discharge accurately and rapidly. However, at present, atmospheric monitoring in China is still reliant on large-scale national monitoring stations, which can provide high-precision monitoring data but are not arranged in large-area and high-density monitoring sites because of cost constraints. As a result, we cannot acquire the monitoring data with sufficiently high spatial precision to locate the discharge sources. Therefore, it is essential that monitoring sites that can acquire data with high space-time accuracy are developed. With the development of the Internet of Things (IoT) [3], the number of IoT devices in general use is continuing to grow rapidly. Application of the IoT to monitor the air quality can provide sufficient data to allow the pollution sources to be located precisely. With characteristics that include on-line monitoring, target tracking, remote maintenance, and an on-line upgrade capability, the IoT can solve the problems of high-space-time-precision air monitoring efficiently by selecting the most appropriate networking mode to realize interconnection and communication for air monitoring equipment. The emerging LoRa (Long Range) [4] is a type of low-power wireless area network (WAN) communication technology that offers several advantages when applied to the IoT. As shown in Figure 1, LoRa applications are becoming more widespread. Using a star network structure, LoRa achieves ultra-low current consumption and provides long-distance transmission via a spread-spectrum modulation design and a reduced communication modulation frequency, and it is resistant to external interference and multipath fading because it uses frequency-hopping spread spectrum technology. These IoT-compatible features make it suitable for device applications that require low power consumption and small data volumes (less than 50 bytes for a single packet). To date, many countries have started to develop applications for the IoT, and some of the required infrastructure, such as LoRa base stations, has been deployed extensively [5].

In addition, a great deal of research is being conducted into the application of LoRa to IoT. For example, in China, Wang et al. realized a power meter reading module through use of LoRa [6], Sun et al. applied LoRa technology to power a communications network [7], and the study of long-distance indoor location functions have also been performed based on LoRa's localizable features [8]. Overseas, Pasolini et al. studied the application of LoRa to smart city projects [9], while Cerchecci et al. conducted research into the use of LoRa for urban domestic garbage collection management [10]. These numerous research areas demonstrate the current development prospects of LoRa. There are also some research results on atmospheric environmental monitoring methods. Olalekan A.M. et al. proposed using networks of low-cost air quality sensors to quantify air quality in urban settings [11]. The study introduced the collection of pollutant concentrations by using general packet radio service (GPRS) network communication, and the measurement results of the electrochemical sensor are analyzed in detail. L. Capezzuto et al. also proposed an interesting monitoring method through mobile phones to provide a way for the citizen to easily obtain his own sensor node and quickly start participating in the air monitoring [12]. M.I. Mead et al. provided evidence for the performance of electrochemical sensors at the parts-per-billion level, and then they outlined results obtained from deployments of networks of sensor nodes in an autonomous, high-density, static network in the wider Cambridge (UK) area [13]. Therefore, it can be seen from the above studies that monitoring of the atmospheric environment is a research hotspot and there is also a large market demand.

Based on the above research, this article introduces a high-time-space-precision atmospheric environmental monitoring system based on LoRa (which is hereafter referred to as "the system"). The system was designed and implemented based on the requirements of field environment monitoring applications. The system is modular in terms of both hardware and software to enable it to meet later expansion requirements. Furthermore, the system was designed with full consideration of aspects such as equipment power consumption, reliability, hardware volume, and actual cost. The actual application also demonstrated that the system has strong market prospects. The design, implementation and application of the proposed system are described in detail below.

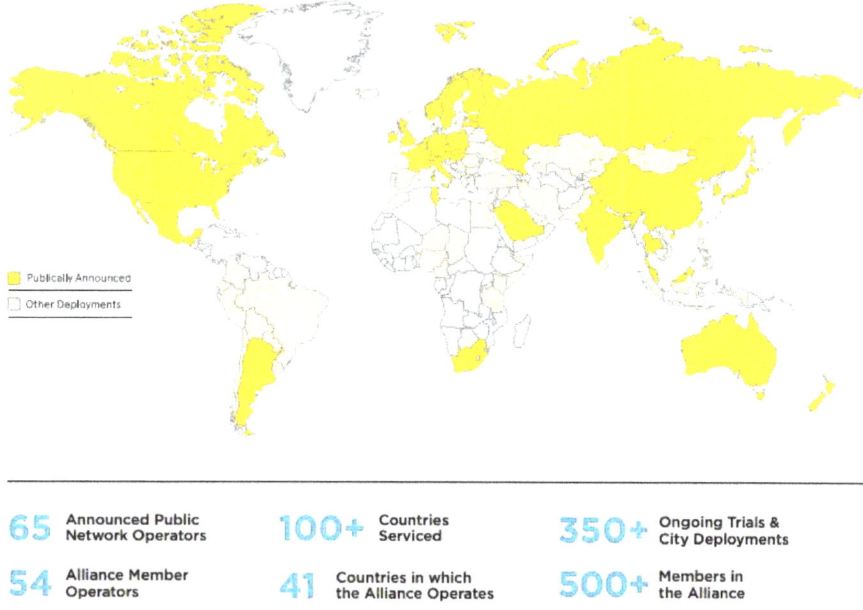

Figure 1. Global distribution of long range (LoRa) usage (which was made by the LoRa alliance and can be found at https://lora-alliance.org/). The dark yellow parts indicate that the application of LoRa technology has been publicly announced, the light-yellow parts indicate that the LoRa technology has a small number of applications, and the light grey parts indicate that the LoRa technology has not yet been applied.

2. System Structure

When monitoring the air quality, it is essential to use a high-density device layout because the environment is usually both complex and volatile [14]. First, the devices must remain sufficiently stable to ensure a long-term monitoring capability. In general, because of the limited access to a mains power supplies in most scenes, the devices are required to be powered by solar energy [15] and they must meet low power consumption standards. Second, it should also be a simple process to replace sensors in the proposed devices to allow the system to collect different types of pollutants. The monitoring equipment that is described in this article is designed in modular form [16], which means that each module's functionality will be realized in accordance with specific application requirements. To meet the requirements for low power consumption and on-line collection of atmospheric pollutants and PM concentrations, the system is largely designed in two parts, i.e., the device side and the server side (as shown in Figure 2). The device consists of the following components: a central control processor module, a sensor data collection module, a data storage module, a power module, and the LoRa communication module. Each module is connected to the processor and performs data collection, storage, and remote transmission tasks via coordination of the software.

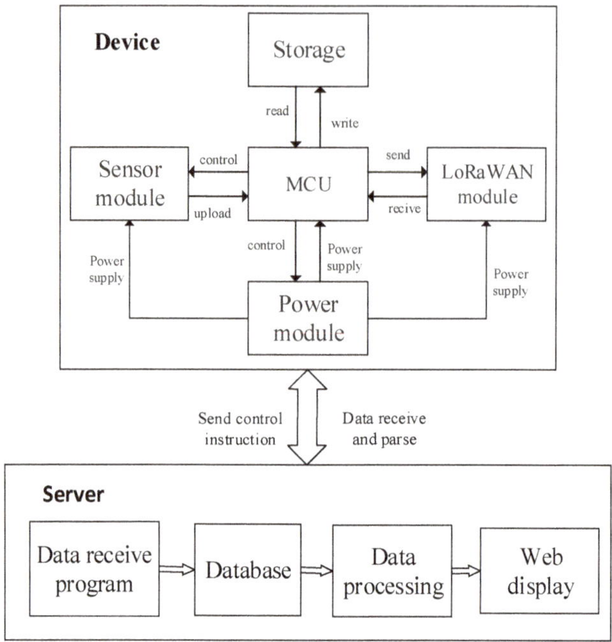

Figure 2. LoRa monitoring system consists of monitoring devices (**top**) and a data server (**bottom**). Arrows between different modules shown data flow and module control procedures. MCU = microcontroller unit; WAN = wireless area network.

2.1. Node Device

The central control processor is mainly based on the STM32F107 [17] chip that is manufactured by STMicroelectronics at Geneva, Switzerland, which is based on a chip from the 32 advanced RISC (Reduced Instruction Set Computer) machine (ARM) Cortex™-M3 series, with a rich array of peripherals and sufficient random-access memory (RAM) space. This module controls the functions of the entire system, including data processing, data storage, and the software drivers required for the communication module and power system control.

The sensor data collection module largely contains electrochemical gas sensors (for SO_2, CO, O_3, and NO_2) and laser scattering sensors (PM2.5 and PM10 particulates). Their specific parameters are shown in Table 1. The sensor interface uses a serial port mode. To ensure that sufficient numbers of serial port sensors can be connected, the system uses a digital switch to poll the sensor data, thus effectively reducing the required number of microcontroller unit (MCU) serial ports.

The data storage module uses a serial peripheral interface flash memory (SPIFLASH) [18] to store the relevant data. There may be some scenarios in which the communication connection is disconnected for a period of time. Therefore, to ensure the continuity of the collected data, it is necessary to store any data that cannot be uploaded in time. At the same time, the system's power consumption can be further reduced by reducing the frequency with which data is uploaded.

The LoRa module controls the wireless data transmission functions of the device. The module is connected to the base station using the LoRaWAN protocol [19]. During periods where there are no data uploads, the module operates in a very low power consumption state. However, when there is an upload, the device is woken up to perform the data upload and receive the downlink data from the base station.

The power module controls the power supply to the hardware system. The power supply uses a combination of solar energy and a battery to ensure that the device can work normally in remote locations where there is no access to a mains supply. In accordance with the formulated power supply strategy, the module will supply power to other modules when required, and the power supplies of these other modules will be turned off if there is no task to be performed, which will save energy to a great extent.

Table 1. The atmospheric sensors' specific parameter characteristics (model, measuring range, measuring principle, resolution, response time, accuracy) used in the system. PM = particulate matter.

Type	Model	Measuring Range	Measuring Principle	Resolution	Response Time (s)	Accuracy
O_3	7NE/O3-5	0–2000 $\mu g/m^3$	Electrochemistry	2.0 $\mu g/m^3$	<30	$\leq \pm 5\%$
SO_2	7NE/SO2-1000	0–2500 $\mu g/m^3$	Electrochemistry	2.5 $\mu g/m^3$	<30	$\leq \pm 5\%$
NO_2	7NE/NO2-1000	0–2000 $\mu g/m^3$	Electrochemistry	2.0 $\mu g/m^3$	<30	$\leq \pm 5\%$
CO	7NE/CO2-1000	0–200 mg/m^3	Electrochemistry	0.2 mg/m^3	<30	$\leq \pm 5\%$
PM2.5	SDS011	0–2000 $\mu g/m^3$	Laser scattering	0.1 $\mu g/m^3$	<10	$\leq \pm 10\%$
PM10	SDS012	0–2000 $\mu g/m^3$	Laser scattering	0.1 $\mu g/m^3$	<10	$\leq \pm 10\%$

2.2. Server Side

The server side is in charge of receiving, passing, storing, and displaying the node device data. The data reception program is based on the Socket communication real-time online monitoring network port. When data are received by the program, these data are parsed and are then written into the MySQL database using the database storage table field. The webpage will then query the database periodically. When there is a new data update, this update will be analyzed with reference to the historical data, and the results will be displayed on the webpage, thus fulfilling the visualization and readability requirements for the monitoring data.

3. System Design

The system was designed using a combination of software and hardware. The software includes a software platform based on the browser/server (B/S) [20] framework and the embedded system software. The hardware is the front-end data acquisition main board. To meet the demand for collection of atmospheric pollutant concentrations, the system uploads the collected data using the LoRaWAN protocol. The system platform design is divided into hardware, software, and system key technology design aspects, and each of these aspects will be introduced in detail below.

3.1. Hardware Design

3.1.1. Main Board

The main board is pictured in Figure 3. As the core component of the atmospheric monitoring node, the main board is composed of a microcontroller, a peripheral circuit, a charge management unit, a power conversion unit, a J-Link component, a serial port setting unit, a data storage unit, a clock management unit, a network interface, and a signal transceiver. Using an STM32F107 chip and the embedded Beijing Normal University Operation System (BNUOS) which developed by the State Key Laboratory of Remote Sensing Science of Beijing Normal University, the microprocessor controls data acquisition in the field environment and data storage, and also manages communication between the collector and the remote server, and communication between the collectors.

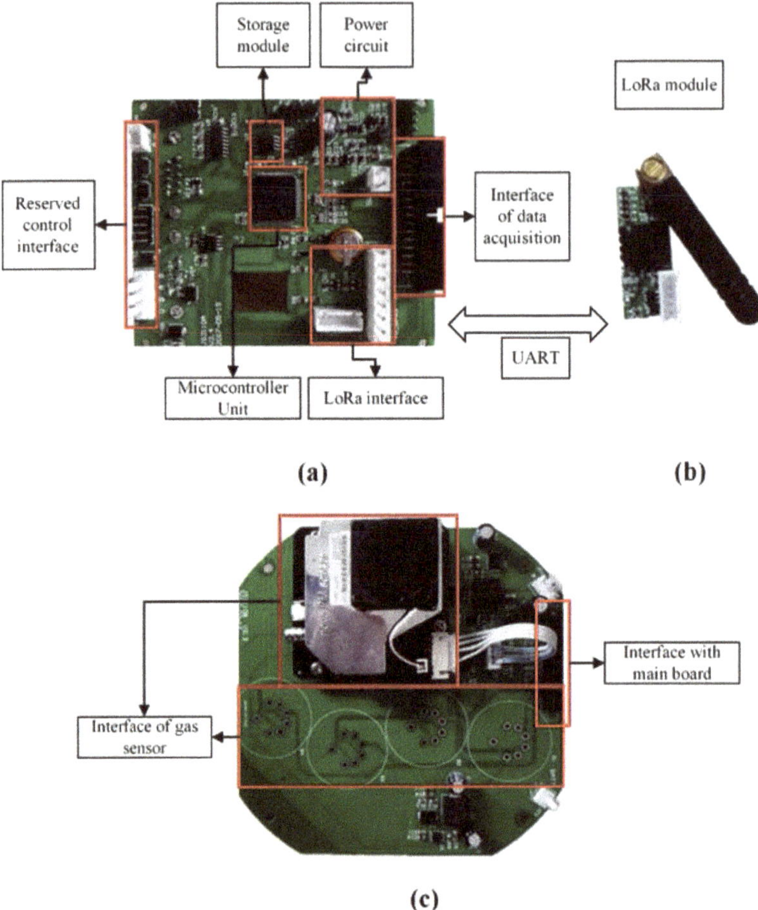

Figure 3. LoRa acquisition equipment hardware structure, consisting of three parts. (**a**) Main board, (**b**) LoRa module, (**c**) Data Collection Board. UART = universal asynchronous receiver/transmitter.

3.1.2. Data Collection Board

Most scientific research requires simultaneous observation of multi-parameter data, which means that a single set of equipment must be used to connect multiple different types of sensors at the same time or to connect equipment from different manufacturers (or even from different countries). To improve the versatility and extensibility of these platforms and support flexible collocation of the different components and devices in a diverse range of projects, it is necessary to design interface modules with increased numbers of functions. Therefore, the main board and the data collection board are separated on these platforms. While the main board is integrated with the serial port, Universal Serial Bus (USB), and Ethernet ports that are common to most current sensors, the data collection board is designed separately to connect each of the sensors. For digital sensors, the data collection board provides a sensor-to-processor communication interface; for analog sensors, if the sensor signal is to be amplified and converted, the data collection board provides the functional components required to link the final signal to the processor; for the output pulse sensor, the data collection board shapes the pulse that is output by the sensor and connects this pulse to the processor. In addition, the data collection board is also both scalable and customizable. The input/output (IO) port on the main board, which is connected to the data collection board, contains 12 analog-to-digital converters (ADCs) and 12 digital-to-analog converters (DACs), a universal synchronous and asynchronous receiver-transmitter (USART), a controller area network (CAN), an inter-integrated circuit (I^2C) bus, a timer, and a counter. In addition to the analog and digital signals, sensors that output other signals such as universal synchronous and asynchronous receiver (USAR), recommended standard 485 (RS485), CAN, I^2C, and serial digital interface at 1200 baud (SDI12) sensors can be connected to the data collection board. Therefore, these sensors can be customized depending on specific demands.

On the platform, the data collection board is as shown in Figure 3; through this board, the sensors can communicate and provide data interaction with the main board. For the different types of sensors described above, the data collection board interacts with the main board in different ways. For example, for the digital sensors, the data collection board only provides an interface to allow these sensors to communicate with the main board. For the analog sensors, however, the data collection board amplifies and converts the electrical signals that are output by the sensors, and then hands them over to the main board via special interfaces for processing.

3.2. Software Design

3.2.1. Software Design of Server

When the central server processes the data, a source analysis model and the weather element analysis environmental quality trends are established in combination with the actual data. The software platform is thus designed based on the demands described above.

The atmospheric environmental monitoring system software is created in the Django framework [21], which stores data using a MySQL database and communicates with the terminal collection device via Socket communication. The terminal application software that receives the device uploads the coded data via a data network, and then analyzes, stores, and displays the data while also providing data visualization and downloading functions for further data analysis. The application software has two main functions: data reception and storage, and real-time data analysis and publication.

The data communication connection between the user server and the device is illustrated in Figure 4. The device is connected to the LoRa base station through the star topology of the LoRaWAN protocol. The LoRa base station accesses the backbone network using the general packet radio service (GPRS), Wi-Fi [22], or Ethernet connections, and completes the network connection to the cloud server through the TCP/IP (Transmission Control Protocol/Internet Protocol) Internet protocol suite [23]. The user server transmits the data to the cloud server via stream sockets, which provide connection-oriented, two-way, and ordered data flow services based on the TCP. The user server application software collects the data that are transmitted by the terminal device via the Socket data receiving stream to the server data pool. The application program then obtains the corresponding data bits based on the data sequence that was transmitted by the device and the corresponding sensor flag bits, and thus acquires the real data using the corresponding conversion factors after the data bits are translated into collection data. The server application program then stores the data in the MySQL relational database. After the receiving program receives the data that are sent by the terminal device and decodes these data, it converts the data into the corresponding data model required for the data call.

The application program visualizes the collected data in accordance with the demands of the users. The overall data visualization framework used here is the popular web framework, Django. The corresponding request link is sent to the background server via the address request of the foreground and the server then obtains the corresponding processing function via the corresponding route mapping. The processing function performs the data processing required, and then the data obtained after processing are fed back to the foreground page for display rendering. Depending on the users' demands, the data analysis device can conduct hourly average, daily average, monthly average, and air quality index (AQI) [24] calculations of these data and the corresponding analysis results are then fed back to the users.

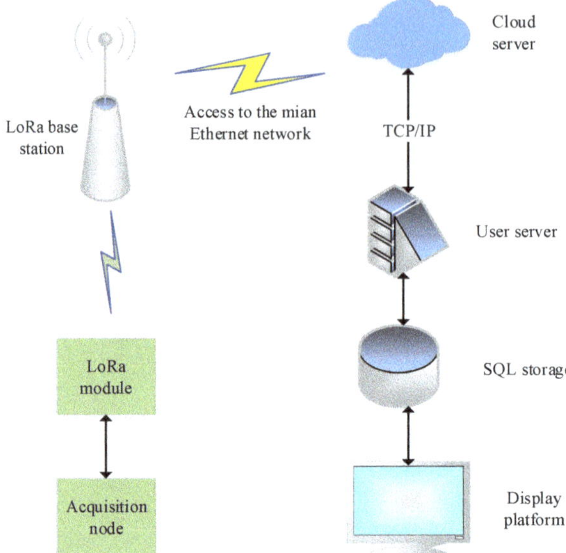

Figure 4. The data communication connection between the user server and the device. TCP/IP = transmission control protocol/internet protocol; SQL = structured query language.

In the system, according to the LoRa base station manufacturer introduction and application results, 1000 collection nodes can be supported under one LoRa base station [25]. The communication rate of LoRa is 292 bps–5.4 kbps, so communication delay is very short. The system has six digital sensors: O_3, SO_2, NO_2, CO, PM2.5, and PM10. In practical application, considering the sensor startup time is <30 s (as shown in Table 1), the acquisition frequency can approach 1 min, and the acquisition frequency can be set by sending control data to the device remotely, so the collection of the main basic parameters of the atmosphere can be completed. The data requirements for environmental governance analysis can be met through these six parameters.

3.2.2. Embedded Software Design

(1)   Main System

The embedded platform software flow is depicted in Figure 5. To reduce the power consumption of the system, the main task module has a system sleep function. Before the system can enter the sleep state, it must determine the tasks that are to be executed after the next system wake-up. The main tasks are as follows: 1. performing the sampling task; 2. opening the LoRa communication module and uploading the sampled data; and 3. opening the LoRa communication module and testing whether or not the link with the server is normal.

In this system, there are several places where the sleep program can be called or the sleep flag can be set. However, the sleep state is only allowed after the sampling and data transmission tasks have been completed or if the battery voltage is below the minimum level required.

In addition, in the case where there is a setup tool or a setup software connection, the system will also wake up automatically.

(2)   Timing Module

The timing module has two different situations. 1. In systems with a sleep function, the timing module provides a trigger to start the sampling process. In these systems, the turn-on times of the data transmission module and the timing module are preferably staggered. 2. In systems that work continuously without a sleep function, the timing module provides semaphores to start the sampling and data transmission modules. In addition, sampling, data transmission, and Global Positioning System (GPS) time calibration should also be controlled via setup tools or using remote networks.

(3) Wireless Transmission

The wireless data transmission module should be formed using standard send-receive functions. The main tasks of this module are listed as follows: 1. Uploading the data collection information and storing this information in an external flash memory; 2. Uploading the signal intensity and the real-time battery voltage of the module while simultaneously uploading the data to enable observation of the current state of the system; and 3. Setting the system parameters and uploading the system status information via remote control. All required correspondences are checked and confirmed.

(4) Data Storage

In general, the system data storage is divided into two parts: 1. the sample data, which are stored in pieces and 2. the system work log information, which includes the information and the status when the system is working, including the times at which the communication module starts and the reasons why the system goes into sleep mode.

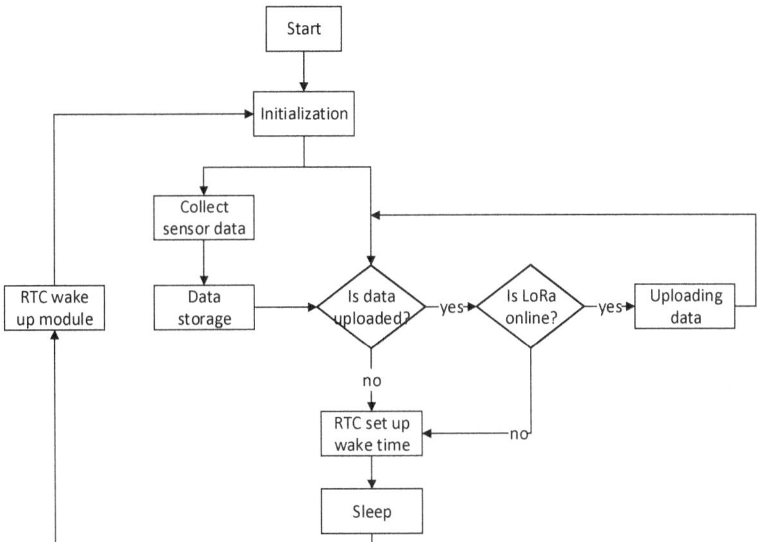

Figure 5. The embedded platform software flow that is applied to the main board of LoRa acquisition equipment. It shows the processing of an acquisition task. RTC = real time clock.

4. Key Technology

4.1. System Operational Stability

Given that these devices commonly operate under the condition that there is no on-site maintenance, self-healing capabilities are strongly required for the device. During the early test procedures, the device halted several times in the unadjusted state and only returned to normal when the device was restarted manually. Usually, these problems are caused by abnormalities in the hardware and software systems. In a mono-chip microcomputer system, data confusion among the various registers and the memory will cause a program pointer error, the pointer not in program area error, and the wrong program instructions, which will interrupt normal operation of the program. As a result, the system that is controlled by this mono-chip microcomputer cannot work normally, meaning that it will stagnate and then finally halt. Therefore, when the system was designed, a hardware watchdog circuit and a software timing reset were installed to ensure that the system can recover spontaneously when any such abnormality occurs. The watchdog is a circuit that checks the internal workings of the chip periodically and issues a hardware reset signal to the chip in the event of an error.

The operational process of the watchdog within the system is as shown in Figure 6. The system contains multiple tasks in the group of Task 1, Task 2 ... Task *n*, and a task monitor which is superior to other monitored tasks. Under the condition that Task 1 to Task *n* all work normally, the Task Monitor clears the hardware watchdog timer within a specified period of time. In the case where any task, e.g., Task *x*, breaks down, the Task Monitor then does not clear the hardware watchdog timer so that the system will reset automatically when the monitored task fails. Additionally, if the Task Monitor itself breaks down, it obviously cannot clear the timer in time, and the watchdog can also reset automatically in that scenario.

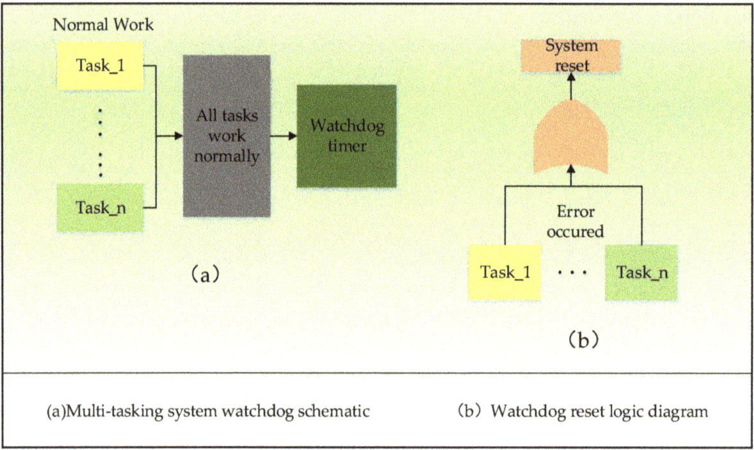

Figure 6. The operational process of the watchdog within the system, which is applied to the main board of the LoRa acquisition equipment in order to ensure the device works normally.

4.2. LoRa Communication Design

The system uses the LoRaWAN protocol chip SX1278 that is manufactured by Semtech [26], which provides long-distance spread spectrum communication based on the standard LoRa modulation technique. The frequency bandwidth of the chip is between 7.8 kHz and 500 kHz. The SF (Spreading Factor) range is from 6 to 12. The available frequency range is from 137 to 525 MHz. To simplify the use of each module in the proposed system, a low-power-consumption chip manufactured by STMicroelectronics, STM32L073xZ, is used specifically as the drive-control chip for the SX1278, with the design structure shown in Figure 7. In this design scheme, the LoRa communication is encapsulated in a transparent transmission module, and the upper application processor transmits and receives the data through the serial port; this not only simplifies the design structure, but also allows the LoRa module to be used in other applications more easily.

The communication process of SX1278 follows the LoRaWAN communication protocol [19]. The LoRaWAN network architecture is a star topology. There are three work modes: Class A, Class B, and Class C, and Class C is used in the system. When LoRaWAN is transmitting, CRC (cyclic redundancy check) is used to ensure sending data correctness. Each LoRa terminal node has its own unique media access control (MAC) address. When the network environment is configured successfully, the sending data includes MAC, channel, and payload. LoRa has a 256 byte data first input first output (FIFO) buffer, which stores the data that is received or will be sent. MCU reads and writes data by accessing FIFO. The receiving and sending flow is shown in Figure 8. In the receiving state, the LoRa module saves the received data in FIFO and carries out CRC checking. After checking successfully, MCU will read the data from FIFO then finish the receiving process. In the sending state, the MCU writes the data into FIFO, and SX1278 automatically sends the data. When the sending is successful, the MCU receives the response message, then the transmission is completed, otherwise the data will be re-transmitted.

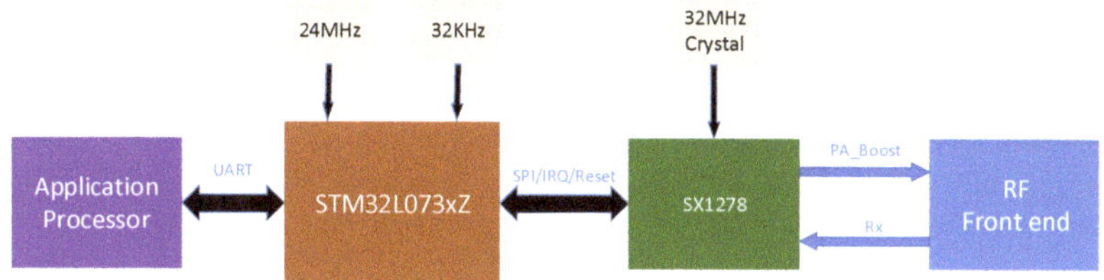

Figure 7. The design hardware structure of LoRa module. SPI = serial peripheral interface; IRQ = interrupt request; PA = power amplifier; Rx = receive; RF = radio frequency.

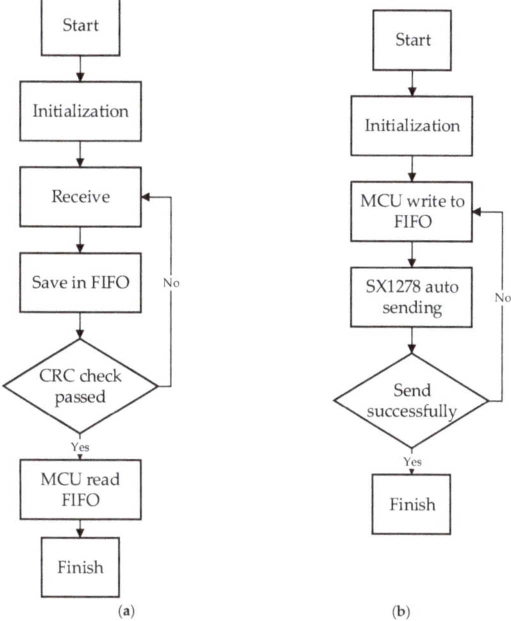

Figure 8. LoRa communication processing between the LoRaWAN chip SX1278 and the MCU of the main board. (**a**) LoRa receiving data. (**b**) LoRa sending data. CRC = cyclic redundancy check; FIFO = first input first output.

5. Application Analysis

To test the device's stability and the accuracy of the collected data, monitoring devices were actually installed at multiple points and operated for long periods to acquire the data. The historical operational behavior of the data analysis devices and the final conclusions are presented as follows.

5.1. Application Stability

The data as shown in Table 2 were collected using several station devices over a period of three months from January to April in 2017 under zero maintenance conditions. The table shows that the data integrity rate of the device is generally above 90%, which indicates that the device did not shut down during this time period. The main reasons for loss of any data are as follows. First, the LoRa communications may be mis-coded or lost during upload, because the distance between the base station and the device will affect LoRa communications. If the device is more than 2 km away from the base station, the data loss rate will then be greatly increased. Second, the device may be abnormal as a result of the effects of the external environment. The watchdog and timing reset strategies are thus applied to ensure that the device can recover spontaneously. It was demonstrated that the system meets the requirements for use in long-term atmospheric environment monitoring.

Table 2. Actual acquisition comparison of multiple station. Theoretical data size (TDS) indicates the volume of data that should be acquired at the normal acquisition frequency within three months. Actual data sizes (ADS) indicates results that are counted by real data. Then, using the equation data acquisition efficiency DAE = ADV/TDV acquired the data integrity rate of the device.

Station	TDS	ADS	DAE
Station-1	2712	2712	100%
Station-2	2712	2559	94.36%
Station-3	2712	2563	94.51%
Station-4	2712	2694	99.34%
Station-5	2712	2692	99.26%
Station-6	2712	2404	88.57%
Station-7	2712	2596	95.72%

5.2. Data Validity

To check the validity of the data that were collected using the device, the data were compared with the authoritative data that were issued by the national control station for the concentration of PM2.5 and CO, which has been causing concern recently, for the whole month. The change trends (as shown in Figures 9 and 10) in the data that were collected using the proposed device based on the LoRaWAN protocol that we introduced here are basically the same as those of the national control station, thus, fully demonstrating that the monitoring and tracking of pollutants can be performed using the proposed system. Practical application tests have shown that the device has full practical application capabilities; its monitoring points can be disposed with high density to meet the application requirements for pollution emission source location under the condition where a large part of the cost must be limited.

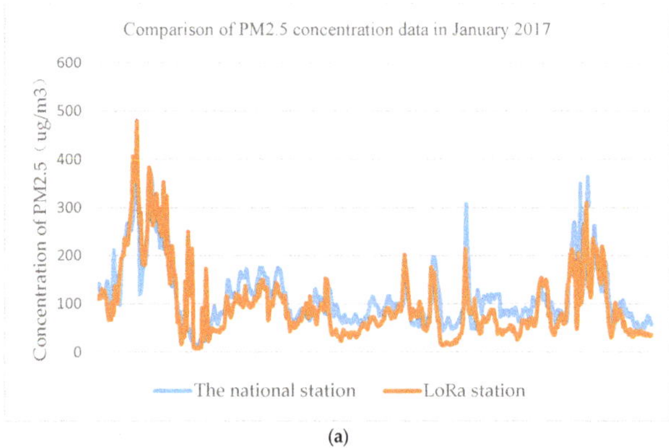

(a)

Figure 9. *Cont.*

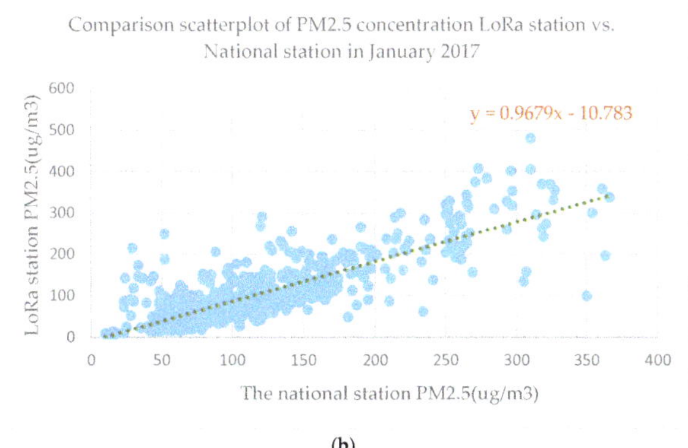

(**b**)

Figure 9. Comparison between the authoritative data that were issued by the national station and the LoRa device station for the concentration of PM2.5 in a month. (**a**) Comparison line graph of the national station data and the LoRa PM2.5 data in a month. (**b**) Comparison scatter plot of the national station PM2.5 data and the LoRa PM2.5 data in month.

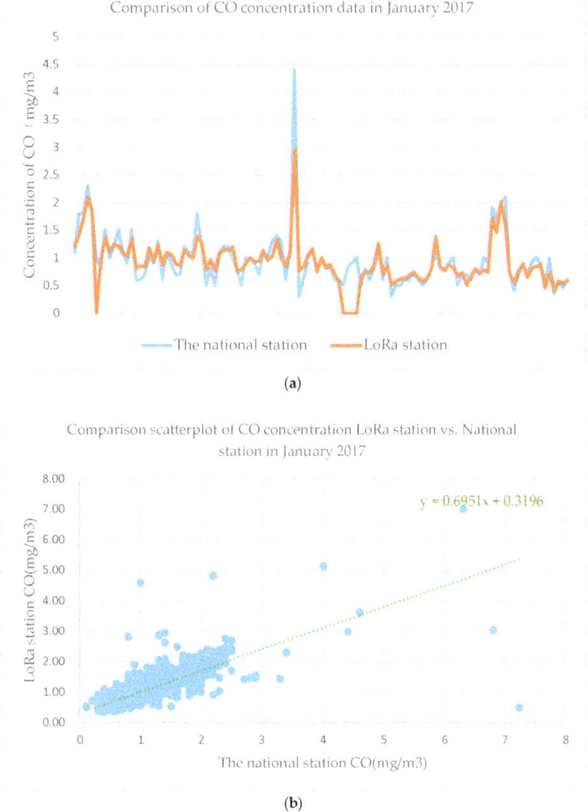

(**a**)

(**b**)

Figure 10. Comparison between the authoritative data that were issued by the national control station and the LoRa device station for the concentration of CO in a month. (**a**) Comparison line graph of the national station data and the LoRa CO data in a month. (**b**) Comparison scatter plot of the national station CO data and the LoRa CO data in month.

5.3. Time Delay

The time delay of the proposed platform system mainly consists of data collection delays and LoRa communication delays. The data collection delay is caused by the start-up times of the sensors. In this system, PM2.5 and PM10 particulates are monitored using laser light scattering sensors [27], which have a start time of less than 10 s. However, the remaining SO_2, NO_2, and other gas pollutants are monitored using electrochemical sensors, which require more than 30 s for startup configuration under the manufacturer's instructions. Additionally, the average time taken by the device to collect the data from the sensors is approximately 32 s. LoRa communication delays are mainly caused by the time taken for LoRa to be registered in the network. Practical applications show that communication delay is less than 5 s when the network signals are strong, while the registration time may be up to 25 s when the signals are weaker. Under the worst-case conditions, the time delay may be as much as 57 s.

5.4. Precision

The system sensors are all digital. Because the sensors that monitor the PM2.5 and PM10 particulates are all designed based on the laser scattering principle, the measurement precision may be affected by foggy weather, during which the tested air would need to be dried for accuracy. The sensors for SO_2, NO_2, O_3, and CO are all based on the electrochemical principle [28], which means that both the temperature and the air pressure easily affect their measurement precision. Therefore, in practice, the data will need to be compensated and calibrated.

5.5. Power Consumption

The system power supply uses a combination of solar energy and battery power. In the practical tests, for operation with low power consumption, the power supply can completely meet the monitoring system requirements. As shown in Figure 11, the battery voltage fluctuation of the device was basically maintained within the 12 V–14.5 V range for the entire year, thus indicating that the device has not experienced insufficient power supply conditions.

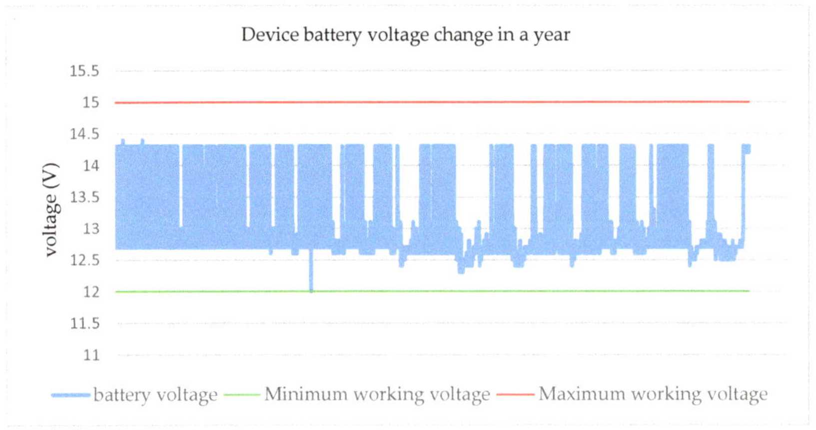

Figure 11. The battery voltage change trend of the device in a year. It shows that the battery voltage (the blue line) was always maintained between the maximum (the red line) and the minimum working voltage (the green line), and the solar-based power supply ensured that the equipment was working properly.

6. Conclusions

Using LoRa communication, the proposed system has performed real-time monitoring of atmospheric pollutant concentrations and the hardware and software design of the platform has been completed. The LoRa module can reduce both the power consumption and the cost of the device efficiently. Furthermore, the combination of the ARM processor with its low power consumption together with the system having a sleep function allows the device to collect data at high frequency for long periods, powered only by solar energy and a battery. The long-term practical tests proved not only the stability and the reliability of the proposed device, but also showed that some of the data may be lost when the distance between the LoRa base station and the nodes is too great or when there are too many obstructions between the base station and the nodes. Therefore, in future work, we will continue to improve the software system to enable the detection and retransmission of lost data, to allow remote upgrading of the device through the server, and to enable adjustment of the sensor parameters. Furthermore, we intend to continue to develop sensors that are compatible with other environmental parameters, such as soil and water parameters, based on the foundation of the system proposed here. Finally, we hope to achieve intelligent environmental monitoring using the advantages of LoRa in the IoT to provide a reliable and stable data reference source for monitoring and management of environmental problems.

Author Contributions: Y.M., L.Z., and X.L. conceived, designed, and performed the main part of the research; R.Y., Y.M., and Q.S. carried out the experiments and analyzed the data; Z.S., and Y.Z. participated in the experiment and contributed to materials and analysis tools.

Funding: This study was supported in part by the Natural Science Foundation of China (Grant No. 41476161), the CERNET Next Generation Internet Technology Innovation Project (Grant No. NGII20160317 and No. NGII20150109).

Conflicts of Interest: The authors declare no conflict of interest.

References

1. Ge, X.; Zhu, T. Study on the Status Quo, Causes and Countermeasures of Urban Atmospheric Environmental Pollution. *Environ. Dev.* **2017**, *29*, 53–54.
2. Thurston, G.D. *Outdoor Air Pollution: Sources, Atmospheric Transport, and Human Health Effects*; Elsevier Inc.: Amsterdam, The Netherlands, 2008.
3. Hu, Y.; Hu, Y.; Yin, B. Internet of Things Information Awareness and Interaction Technology. *J. Comput.* **2012**, *35*, 1147–1163.
4. LoRa Alliance. Available online: https://lora-alliance.org (accessed on 5 February 2018).
5. Peng, W.; Liu, J.; Xin, Z. Research on LoRa Wireless Network Technology and Application Status. *Inf. Commun. Technol.* **2017**, *11*, 65–70.
6. Wang, Y.; Yan, L.; Qin, L.; Li, S.; Wang, F. Research on Power Meter Reading Module Based on LoRa Technology. *Foreign Electron. Meas. Technol.* **2018**, *37*, 46–51.
7. Sun, Y.; Hu, J.; Liu, Y.; Tian, Z. Research on Power Communication Network Based on LoRa. *Inf. Commun.* **2018**, *02*, 218–220.
8. Tangzhou, Y.; Jiang, N. Research on Long Distance Indoor Positioning Based on LoRa. *Comput. Appl. Softw.* **2018**, *35*, 148–154.
9. Pasolini, G.; Buratti, C.; Feltrin, L.; Zabini, F.; De Castro, C.; Verdone, R.; Andrisano, O. Smart City Pilot Projects Using LoRa and IEEE802.15.4 Technologies. *Sensors* **2018**, *18*, 1118.
10. Cerchecci, M.; Luti, F.; Mecocci, A.; Parrino, S.; Peruzzi, G.; Pozzebon, A. A Low Power IoT Sensor Node Architecture for Waste Management Within Smart Cities Context. *Sensors* **2018**, *18*, 1282.
11. Popoola, O.A.; Carruthers, D.; Lad, C.; Bright, V.B.; Mead, M.I.; Stettler, M.E.; Saffell, J.R.; Jones, R.L. Use of networks of low cost air quality sensors to quantify air quality in urban settings. *Atmos. Environ.* **2018**, *194*, 58–70. [CrossRef]
12. Capezzuto, L.; Abbamonte, L.; De Vito, S.; Massera, E.; Formisano, F.; Fattoruso, G.; Di Francia, G.; Buonanno, A. A maker friendly mobile and social sensing approach to urban air quality monitoring. In Proceedings of the SENSORS, Valencia, Spain, 2–5 November 2014; pp. 12–16.
13. Mead, M.I.; Popoola, O.A.M.; Stewart, G.B.; Landshoff, P.; Calleja, M.; Hayes, M.; Baldovi, J.J.; McLeod, M.W.; Hodgson, T.F.; Dicks, J.; et al. The use of electrochemical sensors for monitoring urban air quality in low-cost, high-density networks. *Atmos. Environ.* **2013**, *70*, 186–203. [CrossRef]

14. Feng, Q.; Tao, F. Application of Wireless Sensor Network in Field Measurement. *Electron. Technol. Appl.* **2007**, *09*, 10–12.

15. Sharma, H.; Pal, N.; Kumar, P.; Yadav, A. A control strategy of hybrid solar-wind energy generation system. *Arch. Electr. Eng.* **2017**, *66*, 241–251. [CrossRef]

16. Roth, M.; Hasler, R.; Goblirsch, T.; Franczyk, B. Flexible and Modular Low Power Wireless Networks. *Procedia Comput. Sci.* **2015**, *52*, 695–699. [CrossRef]

17. STMicroelectronics, stm32f107 Datasheet. Available online: https://www.st.com/resource/en/datasheet/stm32f107vc.pdf (accessed on 20 February 2018).

18. Zhen, W.; Li, M.; Shu, J. Flash Storage Technology. *Comput. Res. Dev.* **2010**, *47*, 716–726.

19. LoRa Alliance. LoRaWAN Specification, v1.1[E B/OL]. Available online: https://www.lora-alliance.org/Contact (accessed on 18 January 2018).

20. Wang, S.; Han, C.; Liu, S.; Luo, Q. Establishment on Space Objects Database Management System Using Browser/Server Mode. *Procedia Eng.* **2012**, *29*, 1071–1074. [CrossRef]

21. Xue, Y. *Design and Implementation of Automatic Generation Module Based on Django Framework Management Interface*; Harbin Institute of Technology: Harbin, China, 2014.

22. Muhendra, R.; Rinaldi, A.; Budiman, M. Development of WiFi Mesh Infrastructure for Internet of Things Applications. *Procedia Eng.* **2017**, *170*, 332–337. [CrossRef]

23. Youm, B.; Park, J. Tcp/Ip Protocol Over Ieee-1394 Network for Real-Time Control Applications. *IFAC Proc. Vol.* **2005**, *38*, 37–42. [CrossRef]

24. Ruggieri, M.; Plaia, A. An aggregate AQI: Comparing different standardizations and introducing a variability index. *Sci. Total Environ.* **2012**, *420*, 263–272. [CrossRef] [PubMed]

25. ClaaTek. Available online: http://www.claaiot.com/web/index.php/product/info/61 (accessed on 20 February 2018).

26. Semtech. SX1278 137 MHz to 525 MHz Low Power Long Range Transceiver. Available online: Available: http://www.semtech.com/wireless-rf/rf-transceivers/sx1278/ (accessed on 19 September 2017).

27. Liu, Y. *Research on Suspension Particle Monitoring System Based on Laser Scattering Method*; Harbin University of Science and Technology: Harbin, China, 2017.

28. Pang, X.; Shaw, M.D.; Gillot, S.; Lewis, A.C. The impacts of water vapour and co-pollutants on the performance of electrochemical gas sensors used for air quality monitoring. *Sens. Actuators B Chem.* **2018**, *266*, 674–684. [CrossRef]

Article

A Quantum Ant Colony Multi-Objective Routing Algorithm in WSN and Its Application in a Manufacturing Environment

Fei Li [1,*], Min Liu [2] and Gaowei Xu [2]

[1] Department of Computer Science, Zhejiang University City College, Hangzhou 310015, China
[2] College of Electronics and Information Engineering, Tongji University, Shanghai 201804, China
* Correspondence: lif@zucc.edu.cn; Tel.: +86-153-8112-5261

Received: 5 June 2019; Accepted: 26 July 2019; Published: 29 July 2019

Abstract: In many complex manufacturing environments, the running equipment must be monitored by Wireless Sensor Networks (WSNs), which not only requires WSNs to have long service lifetimes, but also to achieve rapid and high-quality transmission of equipment monitoring data to monitoring centers. Traditional routing algorithms in WSNs, such as Basic Ant-Based Routing (BABR) only require the single shortest path, and the BABR algorithm converges slowly, easily falling into a local optimum and leading to premature stagnation of the algorithm. A new WSN routing algorithm, named the Quantum Ant Colony Multi-Objective Routing (QACMOR) can be used for monitoring in such manufacturing environments by introducing quantum computation and a multi-objective fitness function into the routing research algorithm. Concretely, quantum bits are used to represent the node pheromone, and quantum gates are rotated to update the pheromone of the search path. The factors of energy consumption, transmission delay, and network load-balancing degree of the nodes in the search path act as fitness functions to determine the optimal path. Here, a simulation analysis and actual manufacturing environment verify the QACMOR's improvement in performance.

Keywords: wireless sensor network (WSN); energy; ant colony optimization (ACO); routing algorithm; quantum-inspired evolutionary algorithms

---

1. Introduction

Recent years have seen a worldwide interest in Wireless Sensor Network (WSN) [1] technology, which has been considered one of the most promising technologies in smart manufacturing. Actually, the development tendency of WSN is in accordance with its context of Industry 4.0 [2]. Together with the Industrial Internet, the Internet of Things (IoT) [3], whose kernel is WSN, contributes to the achievement of the connectivity and communication of Cyber-Physical Systems (CPS) [4]. WSN techniques are appropriate for long-term data acquisition for IoT representation in an industrial environment.

WSNs are distinguished from traditional wireless networks by their dissimilar purposes: WSNs are data-centric, while the latter aim for data transmission. In traditional wireless networks, such as Ad hoc and Wireless Local Area Networks (WLANs), the main task is to find the low-latency path between the source node and the destination node, and to improve the utilization of the whole network in order to avoid communication congestion and simultaneously balance network flow. However, in WSNs, a routing method has two main functions: to find the optimal path from the source node to the destination node, and to transmit a data packet along that path. The main aim of network routing improvement is to extend network life and prevent connection errors [5]. The routing method's emphasis is on energy efficiency, because of limited node energy and long lifetime requirements. Meanwhile, since the number of sensor nodes tends to be very large, and these nodes can only obtain local topological information, a suitable route should be chosen by considering local network information.

Since the network is resource- and power-limited, general wireless communication network routing methods are not well-suited for WSNs, especially in industrial fields in which there is demand for high performance in energy efficiency and longevity. Accordingly, some routing approaches have emerged, such as swarm intelligence-based schemes [5,6]. Social insect colonies, such as those of ants and honeybees [7,8], have complex collective behaviors and decentralized management structures, which are similar to parallel, dynamic, and distributed systems. Researchers have studied ant colony optimization (ACO)-based routing schemes to develop high-performance routing methods [9].

In order to improve the limitations of ACO-based routing methods, such as earlier stagnation and slow astringency, this paper considers the idea of using quantum-inspired evolutionary algorithms (QEAs) [10,11] and ACO together, balancing load, real-time transmission, and energy consumption with a multi-objective fitness function. A novel and efficient routing approach for WSNs, called the Quantum Ant Colony Multi-Objective Routing (QACMOR) algorithm, is proposed accordingly. In QACMOR, some quantum computing mechanisms of QEAs, including the quantum bit (qubit) and the quantum rotation gate, are introduced into ACO. The former represents the node's pheromone, and the latter updates it. QEAs are able to avoid premature convergence with a simple implementation, which has more potential for solving large-scale problems than do other general evolutionary algorithms. In multiple objectives, more attention is paid to computation speed by using the look-up table of the rotation angle of QEAs and setting a time-delay factor in fitness function.

The rest of the paper is organized as follows: Section 2 presents the literature review on WSN routing methods. Section 3 sheds light on ACO-based routing in detail. Section 4 explains the proposed QACMOR approach. Section 5 shows the experimental results of performance evaluation and case study validated in a continuous steel casting production line. Finally, Section 6 discusses conclusions and future work.

2. Literature Review

The routing protocol of WSNs should be devised with properties such as energy efficiency, scalability, robustness, and rapid convergence, compared to that of traditional networks. A large number of routing methods have been proposed. Roughly, they can be divided into four categories through the analysis of relevant literature—that is, data-centric, clustering, geographic location-based, and Quality of Service (QoS)-based routing methods.

Data-centric routing was proposed to reduce the flooding overhead caused by transmitting query and data information. In data-centric routing, data request and collection are based on data attributes, rather than only using local interactions [12,13]. Clustering is the most common technique used for achieving energy-efficient and scalable performance in large-scale sensor networks. Cluster formation is a process whereby sensor nodes decide which cluster head they should associate with among multiple clusters [14,15]. The low-energy adaptive clustering hierarchy (LEACH) [15], a typical cluster-based algorithm, divides a sensor network into a set of clusters, through which energy consumption is balanced and reduced. In geographical routing, the physical location of the sensor node is used to guide the path that a packet takes in the network [16,17]. In some cases of WSN application, a higher-communication QoS is demanded, such as reliability and real-time data transmission. The method in [18,19] can be classified in this category.

Routing methods based on swarm intelligence have robust, adaptive, and scalable performance, suitable for autonomous distributed systems [20,21]. Inspired by the foraging principles of honeybees, Saleem et al. [22] proposed a distributed and decentralized routing protocol called the BeeSensor protocol. Camilo et al. [23] studied the application of the ACO metaheuristic to solve the routing problem in WSNs, and came up with an energy-efficient, ant-based routing algorithm (EEABR). Zungeru et al. [24] improved the EEABR algorithm by applying a new scheme to intelligently initialize and update routing tables, reducing the flooding ability of ants for congestion control. In [25], a self-adaptive routing mechanism is presented to ensure reliability and efficiency during data transmission by adopting the dissemination of a pheromone as a model for dealing with dynamic changes in WSN.

QEAs are based on the concept and principles of quantum computing, such as the quantum bit and the superposition of states. As a kind of evolutionary algorithm, a QEA is also characterized by the representation of the individual, the evaluation function, and population dynamics. Learning from the quantum rotation gate strategy of QEAs, Xing et al. [26] introduced an adaptive evolution mechanism for QoS multicasting in IP/DWDM networks, which allowed each chromosome in a population to update itself to a fitter position according to its own situation.

3. Preliminaries

3.1. Energy Consumption Model

Communication is the activity responsible for the bulk of the energy consumption in WSNs [27]. An energy consumption model used in Reference [27] is applied in this study (see Figure 1).

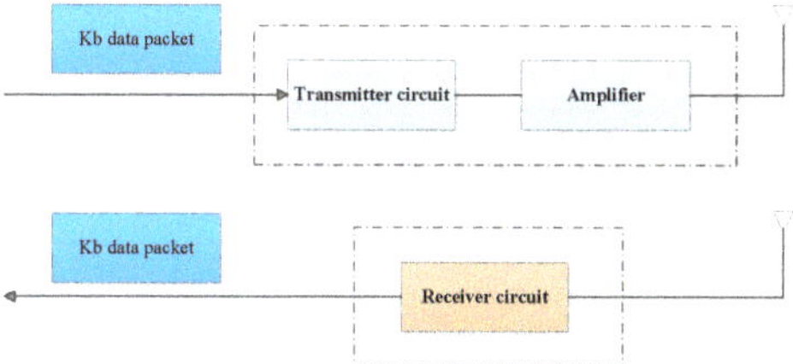

Figure 1. The energy consumption model.

Assumptions are that: the data can reach every node from its neighbors; the data contain information on distance and residual energy; the radio circuit in the sensor has a power control, and can expend the minimum required energy to reach the intended recipients; and radio circuit can be turned off to avoid receiving unintended transmissions. The transmission computation costs and receiving costs for a k-bit message at a certain distance d are shown as follows:

Transmitting

$$E_T(k,d) = E_{elec} \times k + E_{amp} \times k \times d^2 \tag{1}$$

Receiving

$$E_R(k) = E_{elec} \times k + E_{BF} \times k \tag{2}$$

Total energy cost

$$E = E_T + E_R \tag{3}$$

where $E_{elec} = 50nJ/bit$, $E_{amp} = 100pJ/bit/m^2$ for the transmitter amplifier, and $E_{BF} = 5nJ/bit$ when beamforming is used. d represents the distance of two nodes, and k represents the number of message bits.

Thus, by decreasing the communication distance and the volume of data to transmit, energy can be saved.

3.2. Basic Ant-Based Routing (BABR) Algorithm

In ACO, ants exchange data by pheromones, and according to the positive feedback principle, a path with a high density of pheromones has a higher probability of being selected. Such optimization can be adapted to implement basic ant-based routing for WSNs [9,23]:

Step 1: At regular intervals, a forward ant k starts to move from the source node toward the destination. While moving, the identifiers of every visited node are recorded in a list, M_k, and each forward ant avoids traversing a node that has been visited previously.

Step 2: At each node r, a forward ant selects the next hop node in accordance with a certain probability distribution:

$$P_k(r,s) = \begin{cases} \frac{[T(r,s)]^\mu \cdot [E(s)]^\nu}{\sum_{s \notin M_k} [T(r,s)]^\mu \cdot [E(s)]^\nu}, & \text{if } s \notin M_k \\ 0, otherwise \end{cases} \tag{4}$$

where $P_k(r,s)$ is the probability of individual k that moves from node r to node s, and T is the routing table at each node with the amount of pheromone on the link (r, s) stored. E represents the heuristic information given by $1/(C - e_s)$ (C is the initial energy level of the nodes and e_s is the actual energy level of node s), and μ, ν are weight parameters that signify the importance of pheromones versus heuristics.

Step 3: When a forward ant reaches the destination, a backward ant goes back along the links that the forward ant has visited. Before moving, the amount of pheromones that the ant will drop during the trip is computed:

$$\Delta T_k = \frac{1}{N - Fd_k} \tag{5}$$

where N is the total number of nodes, and Fd_k is the distance traveled by the forward ant k.

Step 4: Whenever a node r receives a backward ant from a neighbor node, the routing table is updated:

$$T_k(r,s) = (1 - \rho)T_k(r,s) + \Delta T_k \tag{6}$$

where ρ is a coefficient, and then $(1 - \rho)$ represents the evaporation of pheromones.

Step 5: Once a backward ant returns to the source node, the next interval is continued.

After several iterations, each node will find the best neighbors to which to send a data packet. While the ability and robustness of the ACO-based method qualify it to find a good solution, it still has the possibility of getting stuck in slow astringency and early stagnation.

4. The QACMOR Routing Method

This section first introduces the basic concepts and rules of QEAs, and then elaborates on the QACMOR algorithm for WSNs routing.

4.1. Mechanisms of QEAs

4.1.1. Basic Elements of QEAs

The memory unit in a classical computer is the bit, which only has two states: "0" or "1", whereas the smallest information unit in QEAs is defined as the qubit [10,11]. A qubit could be in the "0" state, the "1" state, or in a linear superposition of both, which is denoted as $\alpha|0\rangle + \beta|1\rangle$, where $|0\rangle$ and $|1\rangle$ represents the quantum state, and a pair of complex numbers (α, β) is defined with $|\alpha|2 + |\beta|2 = 1$, and the value of $|\alpha|^2$ and $|\beta|^2$ indicates the probability of the "0" state and the "1" state, respectively.

A qubit with the size of n can be represented as the following, which has 2^n kinds of states:

$$\left(\begin{array}{c|c|c|c|c} \alpha_1 & \alpha_2 & & \alpha_i & & \alpha_n \\ \beta_1 & \beta_2 & \cdots & \beta_i & \cdots & \beta_n \end{array} \right) \tag{7}$$

For example, a quantum individual with three qubits is given like this:

$$\left(\begin{array}{c|c|c} \frac{1}{\sqrt{2}} & \frac{1}{\sqrt{2}} & \frac{1}{2} \\ \frac{1}{\sqrt{2}} & \frac{-1}{\sqrt{2}} & \frac{\sqrt{3}}{2} \end{array} \right) \tag{8}$$

It can also be represented as:

$$\frac{1}{4}|000\rangle + \frac{\sqrt{3}}{4}|001\rangle - \frac{1}{4}|010\rangle - \frac{\sqrt{3}}{4}|011\rangle + \frac{1}{4}|100\rangle + \frac{\sqrt{3}}{4}|101\rangle - \frac{1}{4}|110\rangle - \frac{\sqrt{3}}{4}|111\rangle \tag{9}$$

which means that the probabilities of the states $|000\rangle$, $|001\rangle$, $|010\rangle$, $|011\rangle$, $|100\rangle$, $|101\rangle$, $|110\rangle$, and $|111\rangle$ are 1/16, 3/16, 1/16, 3/16, 1/16wh, 3/16, and 1/16, separately.

Commonly, $\xi(\xi \subset (-\pi, \pi])$ denotes the phase of the qubit, and the ith bit phase is $\xi_i = \arctan(\beta_i/\alpha_i)$. The position of ξ_i in coordination is given in Figure 2.

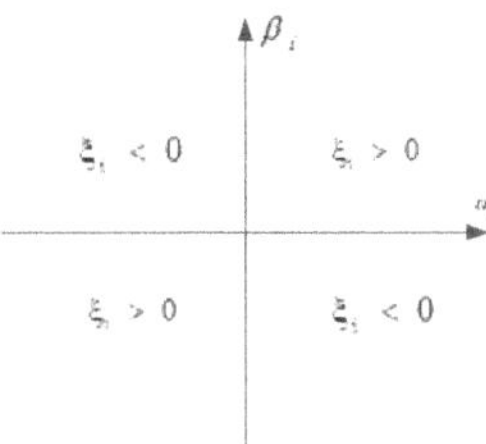

Figure 2. The position of ξ_i in coordination.

4.1.2. The Updating of Qubit in QEAs

In QEAs, the quantum rotation gate updates the qubit. The following formula represents a qubit which rotates θ_i degrees from the original vector, $\left(\alpha_i \quad \beta_i \right)^T$ to $\left(\alpha_i' \quad \beta_i' \right)^T$

$$\begin{bmatrix} \alpha_i' \\ \beta_i' \end{bmatrix} = \begin{bmatrix} \cos(\theta_i) & -\sin(\theta_i) \\ \sin(\theta_i) & \cos(\theta_i) \end{bmatrix} \begin{bmatrix} \alpha_i \\ \beta_i \end{bmatrix} \tag{10}$$

θ_i is the rotation degree according to the following formula:

$$\theta_i = \Delta\theta \times s(\alpha_i, \beta_i) \tag{11}$$

$$\Delta\theta = 5 \times \exp(-t/t_{\max}) \tag{12}$$

In Formulas (11) and (12), $\Delta\theta$ represents the rotation step, controlling the rotation speed; t represents the current number of iterations; and $t_{\max}$ represents the predefined maximal times of calculation determined by the scale of the problem. The function $s(\alpha_i, \beta_i)$ defines the direction:

$$s(\alpha_i, \beta_i) = (d_{ibest}/d_{inow})(\xi_{ibest} - \xi_{inow}) \tag{13}$$

where
$$\begin{aligned} d_{inow} &= \beta_{inow}/\alpha_{inow} \\ d_{ibest} &= \beta_{ibest}/\alpha_{ibest} \\ \xi_{ibest} &= \arctan(\beta_{ibest}/\alpha_{ibest}) \\ \xi_{inow} &= \arctan(\beta_{inow}/\alpha_{inow}) \end{aligned} \tag{14}$$

In Formula (14), α_{inow}, β_{inow}, α_{ibest}, β_{ibest} are the probability of the ith qubit of the current and optimal solution, respectively. Finally, if $s(\alpha_i, \beta_i) < 0$, the θ_i rotates clockwise—otherwise, it rotates counterclockwise.

4.2. The QEAs in QACMOR

4.2.1. Representing the Pheromone with Qubit

In QACMOR, a qubit represents the pheromone for a population with the size of m individuals—that is, $Q = \left(q_1, q_2, \ldots, q_j, \ldots q_m \right)$, $j = 1, 2, \ldots, m$, and

$$q_j = \left(\begin{array}{c|c|c|c} \alpha_1 & \alpha_2 & & \alpha_n \\ \beta_1 & \beta_2 & \cdots & \beta_n \end{array} \right) \tag{15}$$

where n is the number of qubits, $|\alpha_i|_2 + |\beta_i|_2 = 1$, $i = 1, 2, \ldots, n$.

4.2.2. Updating the Pheromone with Quantum Rotation Gate

In QACMOR, similarly to Formulas (10) and (11), the quantum rotation gate G acting on the ith bit of the jth individual q_j of solution Q is described as follows:

$$\begin{bmatrix} \alpha'_{ji} \\ \beta'_{ji} \end{bmatrix} = G \begin{bmatrix} \alpha_{ji} \\ \beta_{ji} \end{bmatrix} \tag{16}$$

$$G = \begin{pmatrix} \cos\theta_{ji} & -\sin\theta_{ji} \\ \sin\theta_{ji} & \cos\theta_{ji} \end{pmatrix} \tag{17}$$

$$\theta_{ji} = \Delta\theta_{ji} \times s(\alpha_{ji}, \beta_{ji}) \tag{18}$$

where $i = 1, 2, \ldots, n$, $\left(\alpha'_{ji}, \beta'_{ji}\right)^T$ represents the updated bit, θ_{ji} is the rotation angle, $\Delta\theta_{ji}$ signifies the magnitude of the rotation angle, and $s(\alpha_{ji}, \beta_{ji})$ is a function of α_{ji} and β_{ji}, and controls the direction of rotation. For the computation speed, the look-up table was applied to compute the rotation angle as shown in Table 1, which includes all feasible solutions. $f(\cdot)$ denotes the fitness function as Formula (20); x_{ji} and b_i represent the ith bit of the jth individual of the current solution x and the best solution b, respectively. The schematic diagram in Figure 3 shows the rotation gate polar plot for a qubit individual.

Table 1. The look-up table of the quantum-inspired evolutionary algorithm (QEA) rotation angle [28].

x_i	b_i	$f(x)>f(b)$	$\Delta\theta_i$	$s(\alpha_i,\beta_i)$			
				$\alpha_i\beta_i>0$	$\alpha_i\beta_i<0$	$\alpha_i=0$	$\beta_i=0$
0	0	False	0	0	0	0	0
0	0	True	0	0	0	0	0
0	1	False	0	0	0	0	0
0	1	True	0.05π	+1	−1	0	±1
1	0	False	0.01π	+1	−1	0	±1
1	0	True	0.025π	−1	+1	±1	0
1	1	False	0.005π	−1	+1	±1	0
1	1	True	0.025π	−1	+1	±1	0

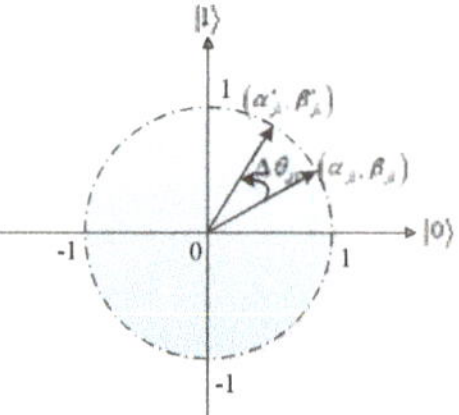

Figure 3. The polar plot of the rotation gate for a qubit individual.

Additionally, a conventional binary solution is significantly important for performance evaluation, and can be obtained by observing the qubits. For example, it is assumed that x_i ($i = 1, 2, \ldots, n$) is a certain bit of the binary individual x, then α_i of the qubit individual is compared with a random number w ($0 < w < 1$). If $|\alpha_i|^2 > w$, then set the value of x_i to be "0", otherwise set the value of x_i to be "1". Therefore, for $Q = \left(q_1, q_2, \ldots, q_j, \ldots q_m\right)$, $j = 1, 2, \ldots, m$, its binary solution is $P = \left(p_1, p_2, \ldots, p_j, \ldots p_m\right)$, while $p_j (j = 1, 2, \ldots, m)$ is a n-length binary individual, and then every element of p_j (for example, p_{ji}) is determined by comparing α_{ji} of q_j with w, $0 < w < 1$.

4.3. The QACMOR Algorithm

The flowchart of the proposed approach is shown in Figure 4. The basic algorithm of QACMOR can be described as follows:

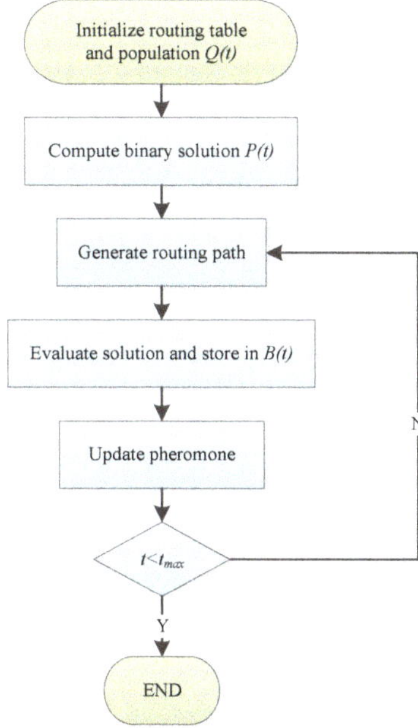

Figure 4. The process of the Quantum Ant Colony Multi-Objective Routing (QACMOR) approach.

Step 1: The initialization step. Add every node and its neighbor nodes into the routing table. A forward ant is generated from source nodes which carry the information of source nodes, sink nodes, and passing nodes. The population is represented as $Q(t) = (q_1^t, q_2^t, \cdots, q_j^t, \cdots, q_m^t)$ with the size of m individuals, where $q_j^t(j = 1, 2, \cdots, m)$ is the jth individual in the tth iteration. The representation is shown as:

$$q_j^t = \left(\left. \begin{array}{c} \alpha_{j1}^t \\ \beta_{j1}^t \end{array} \right| \cdots \left| \begin{array}{c} \alpha_{ji}^t \\ \beta_{ji}^t \end{array} \right| \cdots \left| \begin{array}{c} \alpha_{jn}^t \\ \alpha_{jn}^t \end{array} \right. \right) \tag{19}$$

where n is the number of qubits. Initialize $\alpha_{ji}, \beta_{ji}(i = 1, 2, \cdots, n)$ with $1/\sqrt{2}$. The maximum iterations are represented as t_{max}, and the initial value of the current iterations t is 0.

Step 2: Compute the binary solution $P(t)$. $P(t) = (p_1^t, p_2^t, \cdots, p_j^t, \cdots, p_m^t)$, $p_j^t(j = 1, 2, \cdots, m)$ is a binary individual with n-length. The probable solution is obtained by measurement of $Q(t)$. The value of element p_{ji} in p_j is determined by comparing α_{ji} of q_j with w, $0 < w < 1$.

Step 3: Generate the routing path. Assign m individuals into the source nodes at random. We used the state transition rule to generate the routing path of these individuals. In each step of the decision, an individual positioned on node r moves to the node s in line with Equations (4)–(6).

Step 4: Evaluate the solution and store the best solutions in $B(t)$. The evaluation function of the routing tree is shown as follows:

$$f(t) = \frac{1}{[Z_1(t)]^{C_1} [Z_2(t)]^{C_2} [E_r(t)]^{C_3} [\sigma_r(t)]^{C_4}} \tag{20}$$

$$Z_1(t) = \sum Kd_{rs}^{\lambda}, (r, s) \in Tree(t) \tag{21}$$

$$Z_2(t) = \max Fd_k(t) \tag{22}$$

$$F(t) = \max f(n), (n = 0, 1, \ldots, t) \tag{23}$$

where C_1, C_2, C_3, and C_4 are weight parameters, and $E_r(t)$ and $\sigma_r(t)$ are factors which describe the network load balance, and respectively represent the average value and standard deviation of the load for node r. $Z_1(t)$ is the energy consumption factor, K is an array which indicates the total number of leaf nodes extended from each node in the routing tree, λ is a parameter with a value from 2 to 4 which generally approaches 4, d_{rs} is the distance of link (r,s), $Tree(t)$ denotes the routing tree, $Z_2(t)$ is the time-delay factor, and Fd_k is the distance traveled by the forward ant k.

After the sink node receives forward ant packages, evaluate the solution by Equations (20)–(23), and then save in $B(t)$.

Step 5: Update the pheromone according to the rules of the quantum rotation gate, after receiving back the ant.

Step 6: If the current iterations are less than the maximum iterations, return to Step 3.

It should be noted that QACMOR is an evolutionary algorithm rather than a quantum algorithm, in spite of the fact that the proposed approach is based on quantum computing mechanisms. In QACMOR, some problems in basic ACO can be tackled. The representation of qubit introduces the probability research method, making the balance between exploration and exploitation easier than the conventional ACO algorithm, and adjustment of the magnitude of the rotation angle can make convergence speeds faster. Exploring the unused nodes by using heuristic information, as Formula (4) shows, updating the local pheromone according to Formula (5) and (6) in Step 2, and updating the global pheromone with the quantum rotation gate will generate population diversity, preventing the algorithm from becoming trapped in local convergence or premature stagnation.

5. Experimental Results

5.1. Performance Evaluation

Routing is a crucial process to consider in WSNs when dealing with multiple performance metrics, since routing decisions can impact network lifetime, packet delivery rates, and end-to-end packet delays [29]. Different performance metrics can be used for comparing different routing algorithms in WSNs. The main metrics considered in this paper to validate the performance of the proposed algorithm are as follows:

(1)	General property, such as communication distance, energy consumption, and hops.
(2)	Convergence rate, that is, the number of iterations needed to find an approximation to a fixed point.
(3)	Network lifetime, that is, the duration up to the time when data can no longer be forwarded due to the depletion of energy of the sensor nodes.

Sensor nodes are assigned at random. Figure 5 shows an instance in which the network range was 1000 m^2, and the total number of nodes was 50. Each link between a node and its accessible neighbors was denoted by a dotted line. Figure 5 shows the optimal path obtained by QACMOR, shown as a solid red line. Source nodes were numbered 16, 21, 22, 24, 30, 47, and 50, and the sink node was numbered 1. Notice that the value of t_{max} should be greater than the number of iterations for the algorithm to converge.

Three groups of experiments were conducted on a MATLAB simulation platform. Table 2 lists the values of parameters used in this simulation.

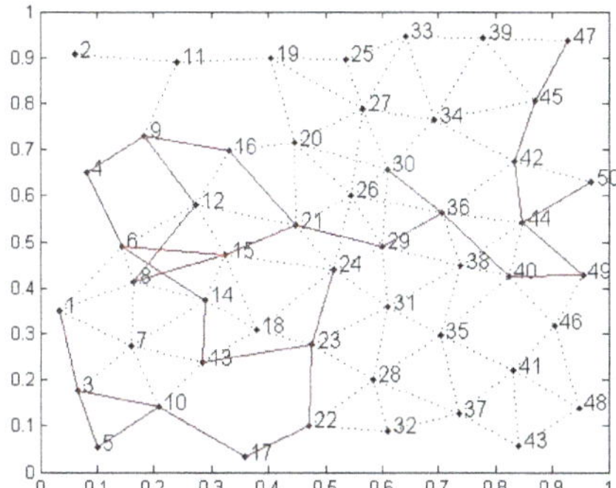

Figure 5. Optimal path in network topology.

Table 2. Parameter-setting in experiments.

Item	Experiment 1	Experiment 2	Experiment 3
Number of nodes	10, 20, 30, ... , 100	50	50
Network range	1000 m^2	1000 m^2	5000 m^2
Initial energy	/	/	0.5 J
C_1	0.5	0.5	0.5
C_2	0.1	0.1	0.1
C_3	0.1	0.1	0.1
C_4	0.1	0.1	0.1
t_{max}	400	400	400

In the first experiment, a comparison of the value of $F(t)$ in cases in which the number of nodes ranges from 10 to 100 was conducted between two algorithms, that is, BABR and QACMOR. The curve lines in Figure 6 show that the values of $F(t)$ for the two algorithms are same at the beginning, and descend as the number of nodes increases. Compared with BABR, the curve of QACMOR has a more sluggish downtrend. The reason for this is that QACMOR takes more properties into account, including energy efficiency, load balance, and time delay.

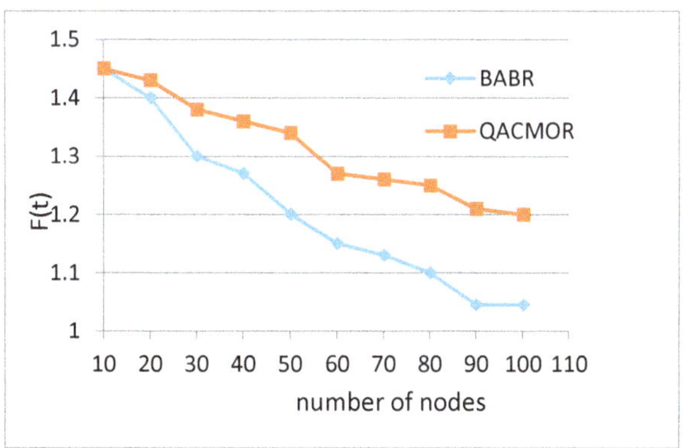

Figure 6. Value of optimal route vs. number of nodes.

The aim of the second experiment was to estimate the convergence property by observing the optimal value $F(t)$ of QACMOR and BABR when the number of iterations grows. As iterations grow, it can be seen in Figure 7 that the value of $F(t)$ tends to be stable. In addition, we notice that QACMOR begins to converge at nearly 200 iterations, while it takes approximately 350 iterations for BABR. This demonstrates that QACMOR has a faster convergence rate than BABR.

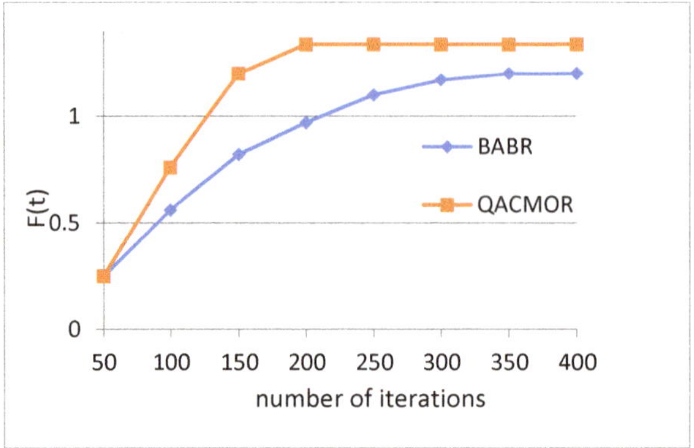

Figure 7. Value of optimal route vs. iterations.

The third experiment evaluates the network lifetime of QACMOR. The experiment is performed on the condition that the number of dead nodes grows. In Figure 8, the x-axis denotes the number of dead nodes, and the y-axis represents the lifetime of the network. It can be seen that the value of lifetime for QACMOR is consistently higher than the same value for BABR, the gap becomes bigger with increasing dead nodes, and maintains a fixed value after 35 nodes, nearly 900 *h*.

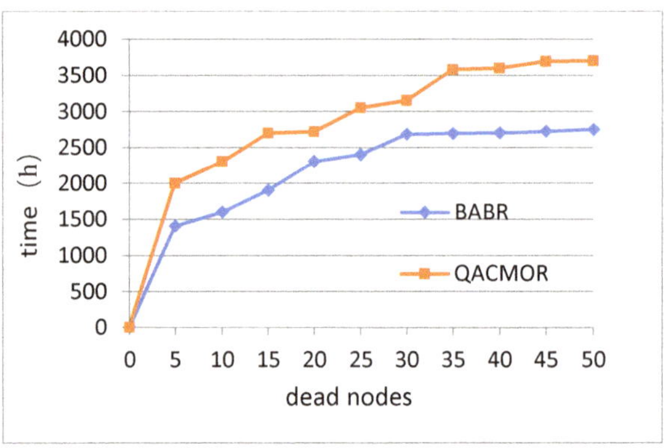

Figure 8. Time vs. number of dead nodes.

The above experimental results indicate that the QACMOR algorithm is capable of use as an efficient and reliable solution for routing, with balanced energy consumption and an improved network lifetime.

5.2. Case Study

In this section, a case study of a maintenance, repair, and overhaul (MRO) system for a steel manufacturing enterprise is illustrated, in order to evaluate the practicality of QACMOR.

Some situations requiring WSNs, such as continuous steel casting lines, present unique characteristics, mainly due to their harsh industrial environment. In the case of a casting line, this is at high temperature and full of powder, dust, and noise. The installation site of the sensor nodes and sink makes it inconvenient to charge or replace the power supply. Therefore, network longevity should be considered. It is important to build routing algorithms which can be adapted to monitor equipment conditions and prolong the WSN lifetime as much as possible. Another major challenge in the harsh environment is insufficient QoS in WSNs, such as delay, bandwidth, and packet loss.

The role of the online monitoring system is to obtain the status information of equipment, including temperature, pressure, and revolving speed. The system architecture is depicted in Figure 9. The complex structure of the continuous casting line made it difficult to install and deploy a reliable cable network, while a WSN had the ability to overcome the field wiring problem. In the WSN, these field data were sent to the Advanced RISC Machines (ARM)-based gateway for data collection, fusion, and processing. Then, the data were sent to the server. At the server, the collected data were imported into a database for further analysis and diagnosis of potential faults by the MRO system.

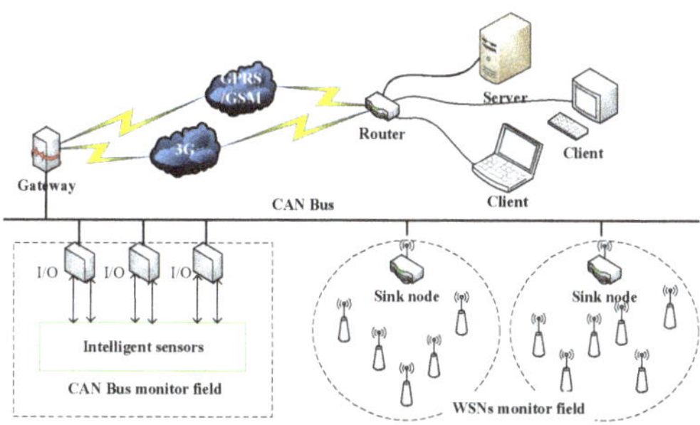

Figure 9. Online monitoring system architecture.

Sensor nodes distributed within the continuous casting line constitute the system's perception layer. Figure 10 shows the installation site of three frame-offset wireless sensors on a segment. In this section, we chose one segment as the test object. Specifically, in one segment, 24 temperature sensors were used to collect information about the working status of the hydro-cylinders; 24 pressure sensors were installed to collect information about the bearings; eight revolving speed sensors were embedded to collect information about the rollers; and three frame-offset wireless sensors were installed onto each segment to monitor the displacement of the segment's frame.

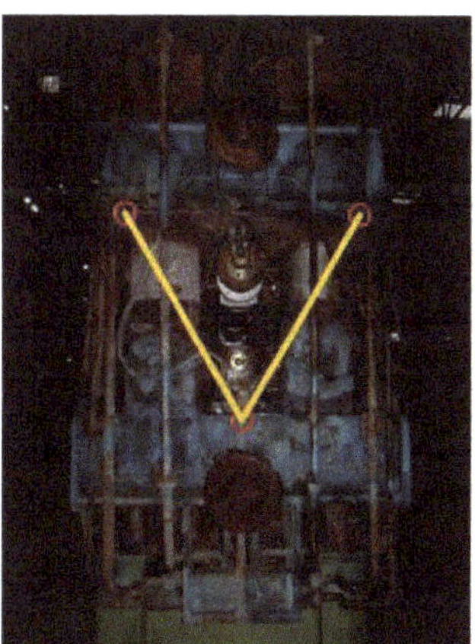

Figure 10. Diagram of three frame-offset wireless sensors.

We conducted a test on one segment in a real casting-shop environment to compare the network lifetime of three algorithms—that is, BABR, AODV, and QACMOR, and verify the running practicality of QACMOR. In this test, the total number of nodes is 60, with one sink node and 59 sensor nodes for one segment. The parameter settings are listed in Table 3, which shows the same weight value (C_1, C_2, C_3, C_4) as that listed in Table 2 in Experiment 3 in Section 5.1. As in that experiment, the comparison of the whole network lifetime was made by observing the number of dead nodes. Results in Figure 11 indicate that in terms of time elapsed before first node death or total network lifetime, QACMOR still has an advantage over BABR, even in harsh working conditions.

Table 3. Parameter-setting in case study.

Item	Value
Number of nodes	60
Network range	300 m × 280 m
Initial energy	0.5 J
C_1	0.5
C_2	0.1
C_3	0.1
C_4	0.1
t_{max}	500

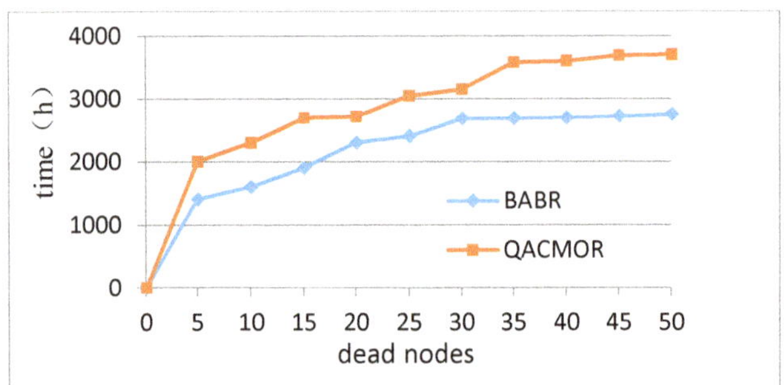

Figure 11. Comparison of network lifetimes in a continuous steel casting line.

6. Conclusions and Future Work

ACO-based routing has been used widely in WSNs. To improve convergence performance and save energy consumption in basic ACO routing methods, quantum computing mechanisms were introduced in the QACMOR method. This paper studied two performance metrics: convergence rate and network lifetime, with reference to the features of industrial continuous steel casting production. Simulation results indicated that the algorithm proposed can rapidly obtain the optimal path with a fast convergence rate, and prolong the network lifetime. A WSN, based on the proposed QACMOR algorithm, was also deployed in an MRO system for a steel manufacturing enterprise. Physical WSN deployment and experiments showed that the proposed QACMOR algorithm is reliable in such applications, after consideration of packet loss based on our previous work [21,30]. In future work, focus and attention should be given to the potential synergies between WSNs and other existing and emerging technologies, such as Cloud Computing and Big Data, so as to improve their overall performance and efficiency.

Author Contributions: F.L. was responsible for methodology, validation and original draft writing, and M.L. guided the whole process, and also reviewed and edited the manuscript, and G.X. worked in software and experimental data collection.

Funding: This research was funded by the Scientific Research Projects of the National Natural Science Foundation of China (Grant no. 61573257), Hangzhou Municipal Science and Technology Bureau of Social Development and Scientific Research Projects (No. 20150533B16), and by the Scientific Research Project of Zhejiang Education Department (No. Y201432791).

Acknowledgments: The research work is partially supported by Hangzhou Key Laboratory for IoT Technology & Application, and Zhejiang Engineering Laboratory Intelligent Plant Factory.

Conflicts of Interest: The authors declare no conflict of interest.

References

1. Xie, X.; Huang, Y.; Niu, C.; Li, C.; Wang, H. Distributed data mining based on deep neural network for wireless sensor network. *Int. J. Distrib. Sens. Netw.* **2015**, *11*, 157453–157457.
2. Li, X.; Li, D.; Wan, J.; Vasilakos, A.V.; Lai, C.-F.; Wang, S. A review of industrial wireless networks in the context of Industry 4.0. *Wirel. Netw.* **2017**, *23*, 1–19. [CrossRef]
3. Distefano, S.; Banerjee, N.; Puliafito, A. Smart objects, infrastructures, and services in the internet of things. *Int. J. Distrib. Sens. Netw.* **2016**, *12*, 8642512. [CrossRef]
4. Liu, K.; Lee, V.C.S.; Ng, J.K.-Y.; Chen, J.; Son, S.H. Temporal data dissemination in vehicular cyber physical systems. *IEEE Trans. Intell. Transp. Syst.* **2014**, *15*, 2419–2431. [CrossRef]
5. Çelik, F.; Zengin, A.; Tuncel, S. A survey on swarm intelligence based routing protocols in wireless sensor networks. *Int. J. Phys. Sci.* **2010**, *5*, 2118–2126.
6. Kang, Q.; Zhou, M.; An, J.; Wu, Q. Swarm intelligence approaches to optimal power flow problem with distributed generator failures in power networks. *IEEE Trans. Autom. Sci. Eng.* **2013**, *10*, 343–353. [CrossRef]
7. Pan, Q.-K. An effective co-evolutionary artificial bee colony algorithm for steelmaking-continuous casting scheduling. *Eur. J. Oper. Res.* **2016**, *250*, 702–714. [CrossRef]
8. Zaheeruddin, D.K.L.; Pathak, A. Energy-aware bee colony approach to extend lifespan of wireless sensor network. *Aust. J. Multi-Discip. Eng.* **2017**, *13*, 1–18.
9. Liu, X. Routing protocols based on ant colony optimization in wireless sensor networks: a survey. *IEEE Access* **2017**, *5*, 26303–26317. [CrossRef]
10. Han, K.H.; Kim, J.H. Quantum-inspired evolutionary algorithms with a new termination criterion, H/sub /spl epsi// gate, and two-phase scheme. *IEEE Trans. Evol. Comput.* **2004**, *8*, 156–169. [CrossRef]
11. Kim, J.-H.; Han, J.-H.; Kim, Y.-H.; Choi, S.-H.; Kim, E.-S. Preference-based solution selection algorithm for evolutionary multiobjective optimization. *IEEE Trans. Evol. Comput.* **2012**, *16*, 20–34. [CrossRef]
12. Albano, M.; Chessa, S.; Nidito, F.; Pelagatti, S. Dealing with nonuniformity in data centric storage for wireless sensor networks. *IEEE Trans. Parallel Distrib. Syst.* **2010**, *22*, 1398–1406. [CrossRef]
13. Chakchouk, N.; Hamdaoui, B.; Frikha, M. WCDS-DCR: an energy-efficient data-centric routing scheme for wireless sensor networks. *Wirel. Commun. Mob. Comput.* **2012**, *12*, 195–205. [CrossRef]
14. Lu, H.; Li, J.; Guizani, M. Secure and Efficient data transmission for cluster-based wireless sensor networks. *IEEE Trans. Parallel Distrib. Syst.* **2014**, *25*, 750–761.
15. Salim, A.; Osamy, W.; Khedr, A.M. IBLEACH: intra-balanced LEACH protocol for wireless sensor networks. *Wirel. Networks* **2014**, *20*, 1515–1525. [CrossRef]
16. Zhang, H.; Shen, H. Energy-efficient beaconless geographic routing in wireless sensor networks. *IEEE Trans. Parallel Distrib. Syst.* **2010**, *21*, 881–896. [CrossRef]
17. Ghaffari, A.; Rahmani, A.M.; Khademzadeh, A. Energy-efficient and QoS-aware geographic routing protocol for wireless sensor networks. *IEICE Electron. Express* **2011**, *8*, 582–588. [CrossRef]
18. Cheng, L.; Niu, J.; Cao, J.; Das, S.K.; Gu, Y. QoS aware geographic opportunistic routing in wireless sensor networks. *IEEE Trans. Parallel Distrib. Syst.* **2014**, *25*, 1864–1875. [CrossRef]
19. Liu, M.; Xu, S.; Sun, S. An agent-assisted QoS-based routing algorithm for wireless sensor networks. *J. Netw. Comput. Appl.* **2012**, *35*, 29–36. [CrossRef]
20. Zungeru, A.M.; Ang, L.-M.; Seng, K.P. Classical and swarm intelligence based routing protocols for wireless sensor networks: A survey and comparison. *J. Netw. Comput. Appl.* **2012**, *35*, 1508–1536. [CrossRef]
21. Wang, J.; Cao, Y.; Li, B.; Kim, H.-J.; Lee, S. Particle swarm optimization based clustering algorithm with mobile sink for WSNs. *Futur. Gener. Comput. Syst.* **2017**, *76*, 452–457. [CrossRef]
22. Saleem, M.; Ullah, I.; Farooq, M. BeeSensor: An energy-efficient and scalable routing protocol for wireless sensor networks. *Inf. Sci.* **2012**, *200*, 38–56. [CrossRef]
23. Camilo, T.; Carreto, C.; Silva, J.S.; Boavida, F. *An Energy-Efficient Ant-Based Routing Algorithm for Wireless Sensor Networks*; Springer Science and Business Media LLC: Berlin, Germany, 2006; Volume 4150, pp. 49–59.
24. Zungeru, A.M.; Seng, K.P.; Ang, L.-M.; Chia, W.C. Energy efficiency performance improvements for ant-based routing algorithm in wireless sensor networks. *J. Sensors* **2013**, *2013*, 1–17. [CrossRef]
25. Saleh, A.M.S.; Ali, B.M.; Mohamad, H.; Rasid, M.F.A.; Ismail, A. RRSEB: A reliable routing scheme for energy-balancing using a self-adaptive method in wireless sensor networks. *TIIS* **2013**, *7*, 1585–1609.

26. Xing, H.; Ji, Y.; Bai, L.; Liu, X.; Qu, Z.; Wang, X. An adaptive-evolution-based quantum-inspired evolutionary algorithm for QoS multicasting in IP/DWDM networks. *Comput. Commun.* **2009**, *32*, 1086–1094. [CrossRef]
27. Heinzelman, W.R.; Chandrakasan, A.; Balakrishnan, H. Energy-efficient communication protocol for wireless microsensor networks. In Proceedings of the 33rd annual Hawaii international conference on system sciences, Maui, HI, USA, 7 January 2000.
28. Wang, L.; Niu, Q.; Fei, M. A novel quantum ant colony optimization algorithm. In *Lecture Notes in Computer Science*; Springer: Berlin/Heidelberg, Germany, 2007; Volume 4688, pp. 277–286.
29. Cobo, L.; Quintero, A.; Pierre, S. Ant-based routing for wireless multimedia sensor networks using multiple qos metrics. *Comput. Netw.* **2007**, *54*, 2991–3010. [CrossRef]
30. Zhang, F.; Liu, M.; Zhou, Z.; Shen, W. An IoT-based online monitoring system for continuous steel casting. *IEEE Internet Things J.* **2016**, *3*, 1355–1363. [CrossRef]

Article

A Posture Recognition Method Based on Indoor Positioning Technology

Xiaoping Huang [1,†], **Fei Wang** [2,†], **Jian Zhang** [3,*], **Zelin Hu** [3] **and Jian Jin** [1]

[1] Institute of Intelligent Machines, Chinese Academy of Sciences, University of Science and Technology of China, Hefei 230031, China; hxping@mail.ustc.edu.cn (X.H.); kljjab@163.com (J.J.)

[2] School of Electronics and Information Engineering, Anhui University, Hefei 230039, China; jswangfei1019@163.com

[3] Institute of Intelligent Machines, Chinese Academy of Sciences, Hefei 230031, China; zlhu@iim.ac.cn

* Correspondence: jzhang@iim.ac.cn; Tel.: +86-0551-65591170

† These authors contributed equally to this work.

Received: 8 January 2019; Accepted: 22 March 2019; Published: 26 March 2019

Abstract: Posture recognition has been widely applied in fields such as physical training, environmental awareness, human-computer-interaction, surveillance system and elderly health care. The traditional methods consist of two main variations: machine vision methods and acceleration sensor methods. The former has the disadvantages of privacy invasion, high cost and complex implementation processes, while the latter has low recognition rate for still postures. A new body posture recognition scheme based on indoor positioning technology is presented in this paper. A single deployed indoor positioning system is constructed by installing wearable receiving tags at key points of the human body. The distance measurement method with ultra-wide band (UWB) radio is applied to position the key points of human body. Posture recognition is implemented by positioning. In the posture recognition algorithm, least square estimation (LSE) method and the improved extended Kalman filtering (iEKF) algorithm are respectively adopted to suppress the noise of the distances measurement and to improve the accuracy of positioning and recognition. The comparison of simulation results with the two methods shows that the improved extended Kalman filtering algorithm is more effective in error performance.

Keywords: posture recognition; indoor positioning; wireless body area network; Kalman filtering; multi-sensor combination

---

1. Introduction

Human posture recognition is an attractive and challenging topic due to its wide range of applications, e.g., smart home environments for the monitoring of physical activity levels, assessment of recovery phases of living independently, and detection of accidental falls in elderly people [1]. Among these applications, the most important one is the elderly health care due to the population aging in the 21st century. According to the US population report, the aged population (over 65 years old) reached more than 50 million in 2017 [2], which represented 15.41% of the US population. China had an aged population of 158 million (10.64% of its total population) in 2017, and it will become one of the most aging countries in the world [3]. Meanwhile, the empty nest ratio of the elderly is rapidly increasing for various reasons. Therefore, health care for the elderly has become a major concern. Posture recognition is one of the key supporting technologies for health care of the elderly.

Traditional human posture detection methods can mainly be divided into two categories: computer vision [3–12] and acceleration sensor data analysis [13–16]. Methods based on acceleration sensors have the disadvantage of complex data processing steps, but their invasion of privacy is well tolerated. Posture detection methods by computer vision technology are mature and have high accuracy in individual posture recognition, but they have the disadvantage of invasion of privacy. Posture parameters extracted from video in [4,5] are skeletal joints and rotation angles. The method performs well in individual human posture recognition with a pan-and-tilt fixed camera, but it suffers difficulties in the recognition of multiple people. Kinect, the Microsoft somatosensory camera, is adopted to extract such parameters as spatial positions of the skeletal joints [6–11] to recognize human posture. Kinect can track at most two bones, six people, 20 joint points in a standing model, or 10 joint points in a sitting model [12]. Due to its poor recognition effect under conditions of multiple participants, Kinect doesn't meet the requirement of multiple people tracking. Both cameras and Kinect have the disadvantage of invasion of privacy, which result in controversies in health monitoring application for the elderly.

On the other hand, human posture detection technology based on acceleration sensor data analysis has been proposed in [13–16]. In [13], a waist-mounted device to detect possible falls of elderly people is presented, and four accelerometer sensors are combined to achieve good performances. However, the algorithm needs to be improved to calculate the optimum thresholds automatically. Musalek [14] used a motion sensor, which is a wearable device capable of wireless communication, to detect the movement of the elderly. Both methods depend on single sensor devices, which provides limited information to recognize human posture. Guo et al. [15] proposed a pose awareness solution for estimating pedestrian walking speeds with the sensors built into smartphones. This method asks the elderly to use a high cost smartphone. Reference [16] presents wearable sensor devices to recognize human postures, and they conducts their experiments with participants wearing three sensors which can reach 90% overall accuracy of human postures. Caroppo et al. [17] described a multi-sensor platform for anomalies, which acquires postures by both ambient and wearable sensors that are a time-of-flight 3D vision sensor, UWB radar sensor and a 3-axis accelerometer. The platform achieves high accuracy in sleep anomaly detection.

Methods based on acceleration sensors can achieve high recognition accuracy in motion states. Since acceleration sensors cannot acquire static information, they have difficulties in the recognition of positions, shapes, and so on. In recent years, the indoor positioning systems have seen increasing development [18–24]. The Microsoft indoor positioning competition has attracted a large number of global teams from both companies and universities each year. The competition has witnessed lots of positioning technologies including wireless local area networks (WLAN) [19], Bluetooth low energy [20], optical light [21], radio frequency identification (RFID) [22], and UWB [23,24]. Among these methods, UWB is considered to be one of the most accurate approaches because it provides positioning estimation with centimeter-level accuracy [25,26].

However, it is doubtful whether the indoor position system can also offer high-accuracy position estimation and posture recognition in dynamic activities. To answer this question, a new posture recognition method with the application of UWB indoor positioning technology is proposed in this paper. In the scheme, a positioning umbrella is designed and constructed first, the UWB distance measuring technology is applied. Receiving tags will be pasted onto key points on the clothes corresponding to human body joints of such as wrists and ankles. Least squares and Kalman filtering algorithms are adopted to reduce the noise interference, thus further improving the positioning and recognition accuracy. Simulation results reveal that the scheme can effectively estimate positions and recognize human postures.

2. Posture Estimation Method

2.1. Design of Positioning System

UWB is one of the hotspots in indoor positioning research. It can achieve centimeter-level accuracy in positioning, and has good multipath resistance performance. The radio can transmit a long distance with low power consumption. Hence, the hardware of the positioning system proposed in the paper is designed based on UWB. The positioning system consists of two parts: a positioning umbrella and receiving tags, as is shown in Figure 1. To facilitate attachment on clothes, the receiving tags are designed to be miniature and hidden. The positioning umbrella is the key component of positioning system. It is made up of core and arms, as shown in Figure 1. The core of the positioning umbrella is composed of a CPU and other control circuit modules such as a communication module, alarm executor module, and location estimation module. The arm of the positioning umbrella connects the UWB radio sender to the central processing unit (CPU), so the senders can work synchronously under the control of the CPU. The UWB module mainly functions as a time of flight (TOF) device for distance measurement.

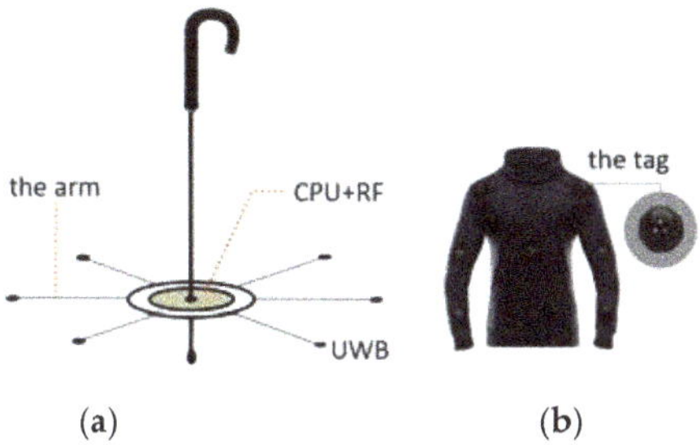

Figure 1. Positioning umbrella and receiving tags. (**a**) Positioning umbrella. (**b**) Receiving tags attached on clothes.

There are several arms in a positioning umbrella system. However, positioning in two-dimensional surface needs at least three arms, and positioning in three-dimensional space requires at least four non-coplanar arms. The more and longer the arms are, the higher the positioning accuracy is [27]. Due to space limitations, the length of arms cannot be infinite. Arms in the system are limited to not more than 1 meter. Meanwhile, the number of arms is limited to eight. Thus, the design can be easy to install and use. As is shown in Figure 1a, the distribution of seven positioning umbrella arms in space can provide position service for any UWB receiving tags in 3-dimensional space.

Powered by a button battery, the receiving tags attached to coats and caps are designed to be very small. In order to ensure low energy consumption, the positioning algorithm should be concise and effective. Positioning algorithms with large computation burdens or long operation time (e.g., particle filters or unscented Kalman filter) do not meet the requirement. In order to make effective use of resources, tags attached to clothes, as is shown in Figure 1b, should be recovered and replaced.

2.2. Position Arrangement of Receiving Tags

The Microsoft Kinect technology collects more than 10 key joints of the human body to achieve posture recognition tasks. Similarly, we construct a human body model with 14 segments and 15 joints, as shown in Figure 2. The system applies 14 receiving tags attached to clothes (e.g., coats trousers and caps) in key joints of human body. By this method, we can easily detect human posture, for instance, we can determine it walking or standing by monitoring the position of receiving tags on knee joints. Also we can decide it falling down or sitting by detecting the position of head tag. For some complex postures such as picking up a cell phone, we need to analyze the combination of tags on hand joint, elbow joint and shoulder. We have named each receive tag and its corresponding position in Table 1. In the next two section (Sections 2.3 and 2.4), the posture recognition method will be discussed based on these 14 receiving tags and their position arrangements.

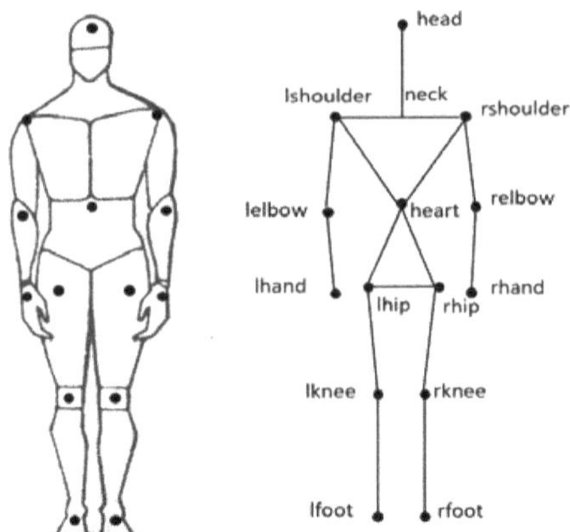

Figure 2. Distribution of UWB receiving tags.

Table 1. Key joints of human body.

Position	Tag Name	Position	Tag Name
Head	head	Central	heart
Left Shoulder	lshoulder	Right Shoulder	rshoulder
Left Elbow	lelbow	Right Elbow	relbow
Left Hand	lhand	Right Hand	rhand
Left Hip	lhip	Right Hip	rhip
Left Knee	lknee	Right Knee	rknee
Left Foot	lfoot	Right Foot	rfoot

2.3. Structure Vector of Human Body

Feature vectors and classifiers are always constructed to implement pattern recognition. Because receiving tags attached to clothes may differ in position for different persons, it is difficult to express the posture characteristics by the absolute position of receiving tags. Therefore, we introduce structure vectors to reproduce the postures of the human body. Due to the different characteristics of the human body, the structure vectors are constructed by the receiving tags which are attached to different positions of clothes to represent trunk, limbs and motion information. Since motion information is represented by a combination of receiving tags, behavior features can be obtained by calculating the vector modulus and vector angle. Compared to the method with one tag, the combination is a more effective method for posture recognition.

By analyzing the characteristics of human body, the vector by connecting receiving tags attached to the key joints is called *structure vector*. Suppose the position of the receiving tag attached to the left elbow is $A(x_1, y_1, z_1)$, and position of the receiving tag on the left hand is $B(x_2, y_2, z_2)$, the vector $\overrightarrow{AB}$ can be expressed as Equation (1):

$$\overrightarrow{AB} = (x_2 - x_1, y_2 - y_1, z_2 - z_1)$$

(1)

Other body structure vectors are similarly constructed. The combination of 14 key joints of Figure 2 in pairs results in 91 structure vectors, among which some vectors are useless to represent human postures. According to the structure characteristics of human body, the vector acquired by two adjacent joints contains the most abundant information. We choose 10 groups of structure vectors that are most capable of expressing changes in body posture, as shown in Table 2.

Table 2. Structure vectors of human body.

Vector Name	Position	Vector Name	Position
$I_{lelbow\text{-}to\text{-}lhand}$	Left Hand	$I_{relbow\text{-}to\text{-}rhand}$	Right Hand
$I_{lshoulder\text{-}to\text{-}lelbow}$	Left Arm	$I_{rshoulder\text{-}to\text{-}relbow}$	Left Arm
$I_{lhip\text{-}to\text{-}lknee}$	Left Leg	$I_{rhip\text{-}to\text{-}rknee}$	Left Leg
$I_{lknee\text{-}to\text{-}lfoot}$	Left Foot	$I_{rknee\text{-}to\text{-}rfoot}$	Left Foot
$I_{head\text{-}to\text{-}lshoulder}$	Head-Left Shoulder	$I_{head\text{-}to\text{-}rshoulder}$	Head-Right Shoulder

2.4. Vector Angle Setting

Besides structure vectors, angle relations between some vectors can effectively reflect the motion information. Figure 3 shows an example for the process of falling down. The angle of $I_{lhip\text{-}to\text{-}heart}$ and $I_{lhip\text{-}to\text{-}lknee}$ is dynamically changing, and the posture angle θ is changing synchronously. Meanwhile, angles can eliminate the structure vector differences originated from shapes and positions of various people.

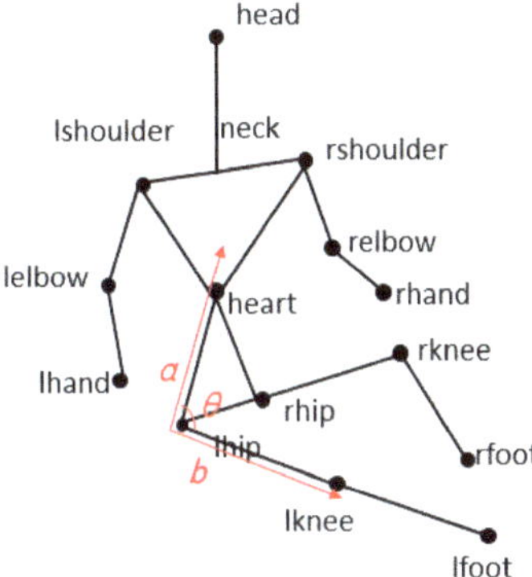

Figure 3. Posture angles of falling down.

The angle of the two structure vectors, $a = (x_1, y_1, z_1)$ and $b = (x_2, y_2, z_2)$, is defined as Equation (2):

$$\langle a, b \rangle = \arccos \frac{a \cdot b}{|a||b|}, \quad |a| \neq 0 \text{ and } |b| \neq 0 \tag{2}$$

If $|a| = 0$ or $|b| = 0$, then $\langle a, b \rangle = 0$, where $a \cdot b$ and $|a|$ are respectively expressed as Equations (3) and (4):

$$a \cdot b = x_1 x_2 + y_1 y_2 + z_1 z_2 \tag{3}$$

$$|a| = \sqrt{x_1^2 + y_1^2 + z_1^2} \tag{4}$$

Now, we make the naming rule of vector angles. The angle of vector $I_{relbow-to-rhand}$ and vector $I_{rshulder-to-relbow}$ is defined as $\theta_{rshoulder-relbow-rhand}$. According to the rule, we select 10 groups of vector angles containing best information that reflects the change of posture, as shown in Table 3.

Table 3. Posture angles of human body.

Posture Angle	Position	Posture Angle	Position
$\theta_{head\text{-}heart\text{-}lshoulder}$	Head-Heart-Left Shoulder	$\theta_{head\text{-}heart\text{-}rshoulder}$	Head-Heart-Right Shoulder
$\theta_{heart\text{-}lshoulder\text{-}lelbow}$	Heart-Left Shoulder-Elbow	$\theta_{heart\text{-}rshoulder\text{-}relbow}$	Heart-Right Shoulder-Elbow
$\theta_{lshoulder\text{-}lelbow\text{-}lhand}$	Left Shoulder-Elbow-Hand	$\theta_{rshoulder\text{-}relbow\text{-}rhand}$	Right Shoulder-Elbow-Hand
$\theta_{lhip\text{-}lknee\text{-}lfoot}$	Left Hip-Knee-foot	$\theta_{rhip\text{-}rknee\text{-}rfoot}$	Right Hip-Knee-foot

3. Positioning Algorithm

3.1. Positioning Principles

According to TOF distance measurement principles of UWB, the angles and length of arms in positioning umbrella are known, and the UWB signal transmitting node at the endpoint of each umbrella arm functions as a base station (or an anchor node). The i-th ($i\in\{1,\ldots M\}$) anchor has its position labeled as $P(x_m^i, y_m^i, z_m^i)$. The anchor position is obtained once the umbrella is constructed. However, the positions of receiving tags are unknown. They will change with body motion. Since receiving tags move with human posture, their positions need to be estimated. Assuming that the position of the j-th ($j\in\{1,\ldots N\}$) tag is $P(x_n^j, y_n^j, z_n^j)$, the distance between the j-th tag and the i-th anchor node can be described in Equation (5), where v is the measurement noise which conforms to Gaussian distribution ($v \sim N(0, R)$), R represents variance of v, and x_n^j, y_n^j and z_n^j are three unknown position parameters. The resolution needs a set of equations with more than four equations similar to Equation (5), as shown in the equation set (6).

$$Y_i^j = \sqrt{\left(x_m^i - x_n^j\right)^2 + \left(y_m^i - y_n^j\right)^2 + \left(z_m^i - z_n^j\right)^2} + v \tag{5}$$

$$\begin{cases} Y_1^j = \sqrt{\left(x_m^1 - x_n^j\right)^2 + \left(y_m^1 - y_n^j\right)^2 + \left(z_m^1 - z_n^j\right)^2} + v_1 \\ Y_2^j = \sqrt{\left(x_m^2 - x_n^j\right)^2 + \left(y_m^2 - y_n^j\right)^2 + \left(z_m^2 - z_n^j\right)^2} + v_2 \\ \qquad\qquad\qquad\vdots \\ Y_M^j = \sqrt{\left(x_m^M - x_n^j\right)^2 + \left(y_m^M - y_n^j\right)^2 + \left(z_m^M - z_n^j\right)^2} + v_M \end{cases} \tag{6}$$

Since there are seven arms in the positioning umbrella in Figure 2, the equation set (6) has seven equations. Linear processing of equation set (6) can be made before the three unknown parameters, and x_n^j, y_n^j and z_n^j are resolved by least squares estimation. Once the positions of receiving tags are solved, the structure vectors and angles can also be estimated.

3.2. Improved and Extended Kalman Filtering Algorithm

In Equation (5), UWB distance measurement is affected by noise v, which has great influence on positioning accuracy. Therefore, an effective algorithm to suppress the measurement noise must be introduced. Since the algorithm needs to be transplanted to a microprocessor, the data processing methods in the receiving tags can't be too complicated for saving energy. Therefore, concise and effective methods such as least square algorithms and extended Kalman filtering algorithms are adopted in our application.

The position of each receiving tag is constructed as $X_n^j(k) = [x_n^j, y_n^j, z_n^j]$, which is called the state variable. Then state equation of these dynamic receiving tags can be represented as the following equation:

$$X_n^j(k+1) = \Phi X_n^j(k) + \Gamma W_n^j(k)$$

$$\Phi = \begin{bmatrix} 1 & 0 & 0 \\ 0 & 1 & 0 \\ 0 & 0 & 1 \end{bmatrix}, \Gamma = \begin{bmatrix} 1 \\ 1 \\ 1 \end{bmatrix} \tag{7}$$

where Φ refers to state-driven matrix, Γ stands for noised-driven matrix, and W is white noise with mean value of 0 and the variance of Q ($W \sim N(0, Q)$). The observation equation is defined in the following form:

$$Y_i^j(k) = h(X_n^j(k)) + v(k) \tag{8}$$

where $h(X_n^j(k)) = \sqrt{\left(x_m^i - x_n^j(k)\right)^2 + \left(y_m^i - y_n^j(k)\right)^2 + \left(z_m^i - z_n^j(k)\right)^2}$.

Then, extended Kalman filter is adopted to process the distance measurement noise. The detailed processing steps are described as follows:

(1) State Prediction:

$$\hat{X}_n^j(k+1|k) = \Phi \hat{X}_n^j(k|k) \tag{9}$$

(2) Covariance Prediction:

$$P(k+1|k) = \Phi P(k|k)\Phi^T + Q(k+1) \tag{10}$$

(3) Kalman Gain Calculation:

$$K = P(k+1|k)H^T[HP(k+1|k)H^T + R(k+1)]^{-1} \tag{11}$$

The Jacobian matrix derived from Equation (11) is shown in Equation (12):

$$H = \frac{\partial h}{\partial X} = \begin{bmatrix} \frac{\partial h}{\partial x_n^j(k)} & \frac{\partial h}{\partial y_n^j(k)} & \frac{\partial h}{\partial z_n^j(k)} \end{bmatrix} \tag{12}$$

(4) Status updating:

$$\hat{X}_n^j(k+1|k+1) = \hat{X}_n^j(k+1|k) + Ke$$
$$e = (Y_i^j(k+1) - h(\hat{X}_n^j(k+1|k))) \tag{13}$$

(5) Covariance updating:

$$P(k+1) = [I - KH]P(k+1|k) \tag{14}$$

The initial filtering value $X(0) = E\{\hat{X}(0)\}$, and the initial variance matrix $P(0) = \text{var}\{\hat{X}(0)\}$. In the filter, e is the Kalman gain, which is calculated from historical data and the latest observation. Too much historical data will lead to cumulative errors. In order to avoid the accumulation of errors, many algorithms such as those described in [28,29] are improved through variable forgetting factors, but it is difficult for them to confirm the values. Therefore, we introduce the rectangular window function to improve the Kalman filter in this paper, as shown in the following form:

$$Y_i^j(k) = \begin{cases} f(Y_i^j(l)), & N - k < l \le k \\ 0, other \end{cases} \tag{15}$$

where $f(Y_i^j(l)) = a_0 + a_1x + \cdots + a_kx^k$, and N is the length of window function with its range $10 \le N \le 30$. Another key improvement for the extended Kalman filter is that the polynomial fitting for the N latest observation Y_i^j is employed. The fitting equation is shown in the following form:

$$A = (X^TX)^{-1}X^TY \tag{16}$$

where $X = \begin{bmatrix} 1 & x_1 & \cdots & x_1^k \\ 1 & x_2 & \cdots & x_2^k \\ \vdots & \vdots & \ddots & \vdots \\ 1 & x_n & \cdots & x_n^k \end{bmatrix}$, $A = \begin{bmatrix} a_0 \\ a_1 \\ \vdots \\ a_k \end{bmatrix}$, $Y = \begin{bmatrix} y_{k-N+1} \\ y_{k-N} \\ \vdots \\ y_k \end{bmatrix}$.

Equation (15) is easy to solve with the coefficient matrix A in Equation (16). Introduction of the window function in the improved algorithm aims at dropping historical data. The smoothed distance obtained by least square polynomial fitting with the latest N observation can help to reduce cumulative errors, and finally to improve the performance of Kalman filter.

4. Experiment

A sketch map of indoor environment is drawn in Figure 4a. The indoor playground is a square with a width of 10 m and length of 10 m. The positioning umbrella is suspended from the ceiling. It is suspended 5 m above the ground. There are seven arms with the length of 1 m. Among the seven arms, six are deployed in the *x-y* plane, and one is deployed in the *z*-direction. When the character moves under the umbrella, we capture the position of the receiving tag of the left foot (other parts are also possible) and draw the trajectory in Figure 4b, which shows the motion curve of the left foot when walking in a straight line at a constant speed. Human motion data for the simulation experiment is obtained from the Unreal Engine 4.0, which is a virtual reality software released in the USA. In this paper, we employ the engine to generate human posture data which are the positions of 14 tags when the character is walking.

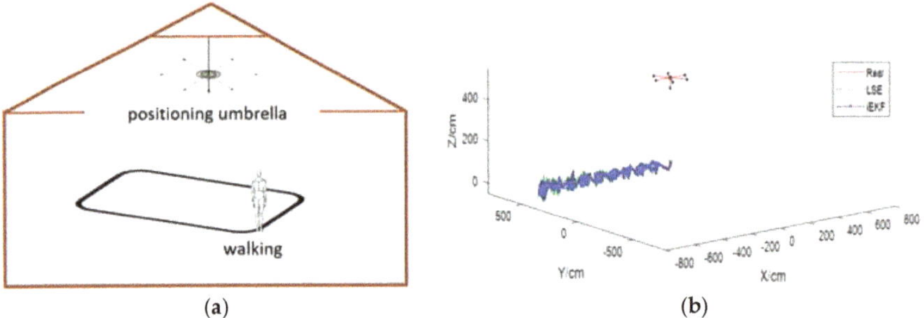

(a)   (b)

Figure 4. Left foot trajectory in walking model. (**a**) Walking model. (**b**) Trajectory of left foot.

In the simulation experiment, the character walks several meters in a straight line at a constant speed. We set the sampling frequency as 119 fps, which means the positions of 14 receiving tags, such as head, shoulders, hands, knees, feet, and so on, should be recorded 119 times per second. Due to the high sampling frequency, we can see the details in the curve from Figure 4b. Then we employ our improved Kalman filtering (iEKF) algorithm. In Figure 4b, "Real" refers to data generated by Unreal Engine 4.0, "LSE" stands for results of least squares estimation (LSE) which is widely used in the field of indoor positioning technology, and "iEKF" represents the results of the improved extended Kalman filtering algorithm in Section 3.2.

5. Results Analysis and Discussion

While in Figure 4b, it is not obvious to decide which method is better. Both algorithms show good tracking effect in three-dimensional trajectories. For further comparison of the performances of two algorithms, we define deviation using Equation (17). Also the mean deviation of each algorithm is defined as Equation (18), where X_{real} are the positions, structure vectors or angles, and $X_{estimate}$ are the estimations by iEKF and LSE methods. The following sections discuss the results of the experiment.

$$deviation = |X_{estimate} - X_{real}| \tag{17}$$

$$mean - deviation = \frac{1}{N}\sum_{k=1}^{N} |X_{estimate}(k) - X_{real}(k)| \tag{18}$$

For further comparison, the mean deviation is an effective metric. The algorithm designed in Section 3 computes the positions of the tags, the vector norms and posture angles by applying iEKF or LSE method. We select two items, respectively, from Tables 1–3. The results of mean deviation are given in Table 4 for the LSE and iEKF methods, respectively. We can conclude that the iEKF is more effective because all the deviations are smaller than with the LSE method.

Table 4. Mean deviation comparison in tag position, vector norm and posture angle.

Name	LSE	iEKF
head	11.70 cm	7.16 cm
left shoulder	11.83 cm	7.24 cm
$l_{heart\text{-}to\text{-}lshoulder}$	0.32 cm	0.20 cm
$l_{lshoulder\text{-}to\text{-}lelbow}$	0.22 cm	0.14 cm
$\theta_{head\text{-}heart\text{-}lshoulder}$	0.0075 rad	0.0046 rad
$\theta_{heart\text{-}lshoulder\text{-}lelbow}$	0.0104 rad	0.0073 rad

(a) Positioning comparison

As described in Section 2.1, we choose 14 receiving tags attached on clothes corresponding to joints of the human body. The positions of the 14 key joints are easily obtained by our methods. In Figure 5, there are two groups of detection results. The first group are the 14 key joints which are detected by the indoor position system, and the others are detected by Microsoft Kinect technology [30].

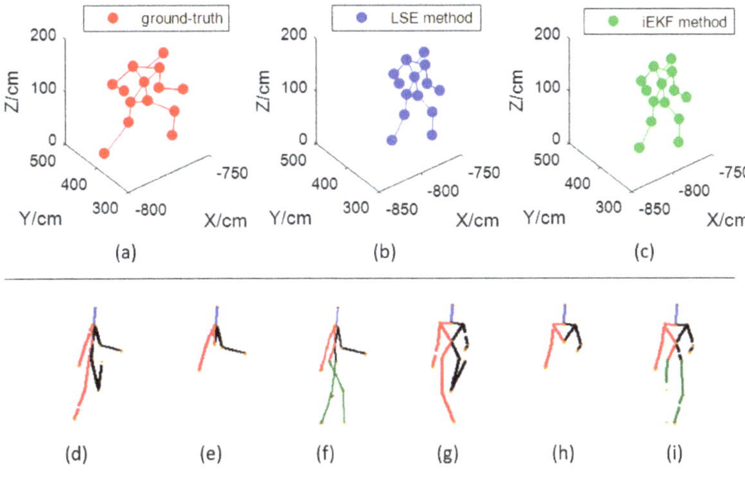

Figure 5. Detecting result for 14 key joints in walking model. (**a**), (**b**) and (**c**) are the first group detected by our designed indoor position system. (**a**) is the original data generated by UE 4.0, (**b**) is the 14 key joints detected by LSE method, and (**c**) is the result detected by iEKF. Figures from (**d**) to (**i**) are the results detected by Kinect sensor. (**d**) and (**g**) are the expected results. (**e**) and (**h**) miss some joints due to shelter or failed detection. (**f**) and (**i**) are the repaired results in literature [30].

The disadvantage of Kinect detection results is that they may be sheltered by something or some joints are missed. Moreover, Kinect cannot output the positions of the 14 key joints. However, these disadvantages are overcome by the indoor position system. The joints are missed only when receiving tags are out of power or broken in the proposal system. Unlike the Kinect method, the receiving tags can calculate the positions.

The positions of 14 joints of human body can be estimated by the iEKF and LSE methods. Taking the head tag for example, Figure 6 shows the deviation of estimation $X_{estimate}$ from real value X_{real} obtained by the two algorithms for head tag. It reveals that iEKF has less deviation and better performance than the LSE method. The mean deviation by the iEKF method is 7.16 cm, lower than the LSE method one of 11.70 cm, so iEKF displays better position accuracy than the LSE method.

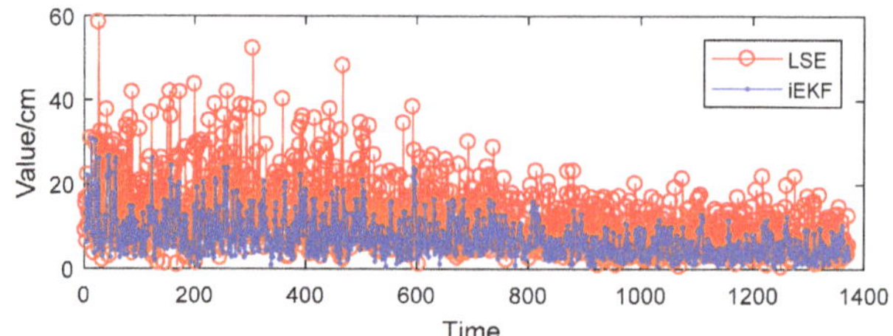

Figure 6. Deviation performances of the two algorithms for head tag.

(b) Struction vector comparison

We now perform further analysis to examine the deviation of structure vectors for body postures. Taking vector $I_{heart\text{-}to\text{-}lshoulder}$ for example, in Figure 7, the red solid line represents the real value of distance between head and left shoulder, the green dotted line refers to the norm of vector $I_{heart\text{-}to\text{-}lshoulder}$ estimated by the LSE method, and the blue dashed line represents the norm estimated by iEKF. Figure 7 shows that the norm of the vector fluctuates between 30.5 cm and 32.5 cm, which is due to the body tilting from left to right when the character is walking. The estimation values of both LSE and iEKF fluctuate near the true value. Figure 8 reveals that iEKF has a smaller deviation and is closer to the true value. The mean deviation between the real value and iEKF is 0.20 cm, which is lower than that of LSE method with 0.32 cm. Thus iEKF has better accuracy.

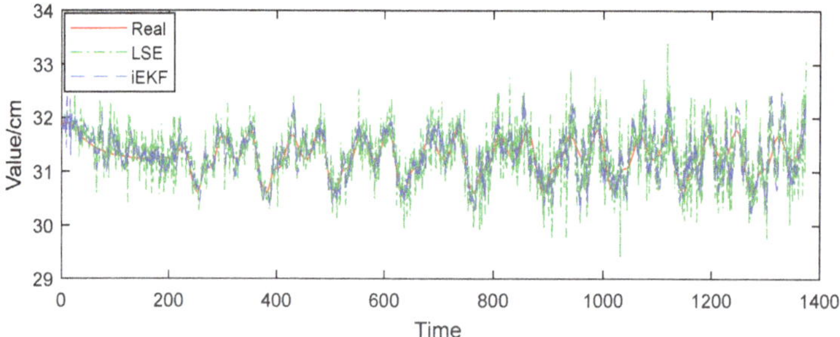

Figure 7. Norm comparison of vector $I_{heart\text{-}to\text{-}lshoulder}$.

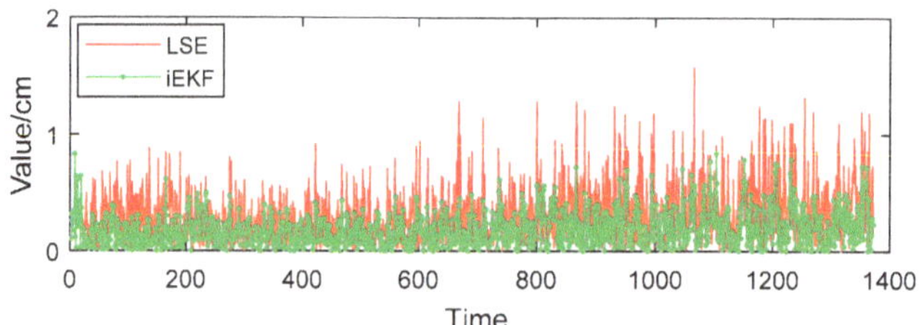

Figure 8. Deviation comparison of vector $I_{heart\text{-}to\text{-}lshoulder}$.

(c) Vector angle comparison

Vector angle reflects well the postures of the human body. Posture angles analysis play an important role in human posture recognition. Vector angle $\theta_{heart\text{-}lshoulder\text{-}lelbow}$ is made up by vector $I_{lshoulder\text{-}to\text{-}lelbow}$ and vector $I_{heart\text{-}to\text{-}lshoulder}$. The angle usually happens under such a condition when people pick up a phone or take a glass of water to drink. In Figure 9, the red solid line, green dotted line and blue dashed line respectively represent the real angle, estimation by LSE and estimation by iEKF,. Also the estimation values of both LSE and iEKF fluctuate near the true value. However, Figure 10 indicates that the deviations of iEKF are lower than those of LSE. The mean deviation estimated by LSE is 0.0104 rad, and the deviation by iEKF is 0.0073 rad which is lower than the LSE method. The iEKF still has obvious advantages in the comparison of vector angles.

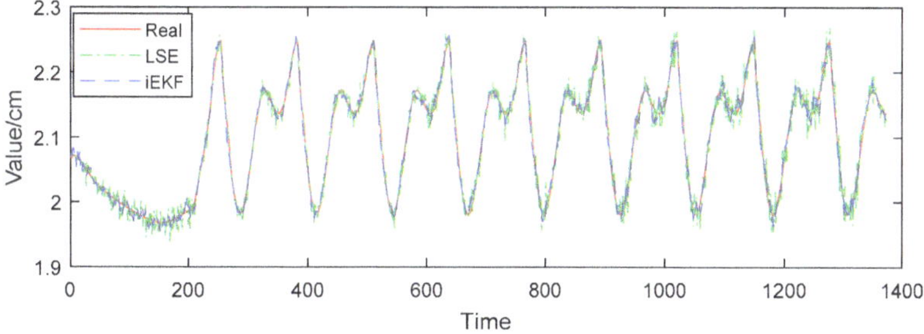

Figure 9. Comparison of two algorithms for $\theta_{heart\text{-}lshoulder\text{-}lelbow}$.

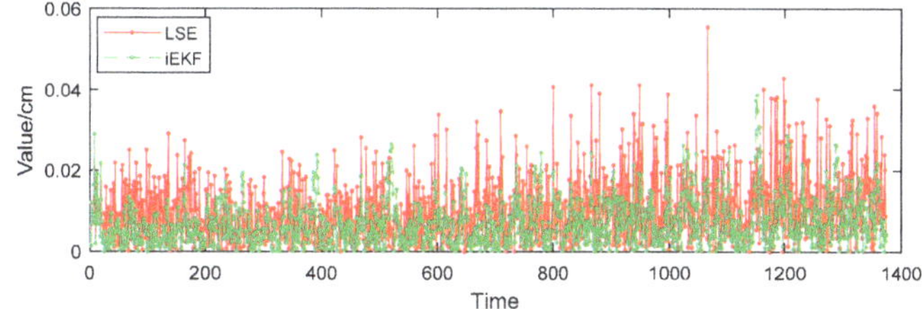

Figure 10. Deviation comparison for angle $\theta_{heart\text{-}lshoulder\text{-}lelbow}$.

(d) Influence discussion for window length N

We refer to the improvement of iEKF in Section 3.2. Now we discuss the influence of the length of N for the window function. When N is set from 0 to 40, we carry out the experiments and calculate the mean deviation of vector $I_{lshoulder\text{-}to\text{-}lelbow}$. The results are shown in Figure 11. From the figure, we can draw the conclusion that optimal performance is achieved when N is between 20 and 25. When N is equal to 23, the deviation is lowest. This doesn't mean that it's globally optimal when N is 23. The optimum N will float slightly for other vectors. Therefore, the ideal choice is $18 \leq N \leq 28$ according to the simulation results.

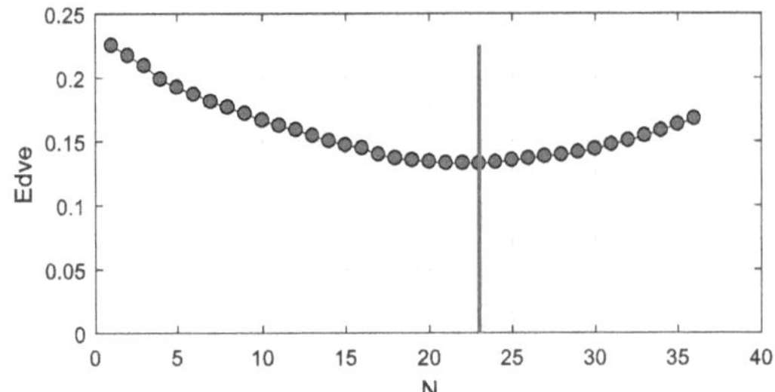

Figure 11. Influence of N for mean deviation of vector $I_{lshoulder\text{-}to\text{-}lelbow}$.

(e) Future work

The outputs of the experiment will be two import products: one is the indoor position umbrella which can provide location services, and the other is the specific clothes with receiving tags which can recognize the postures of the human body. In the near future, we need to carry out the following three tasks before we can convert the experimental results into products:

(1) We have done basic posture recognition work for a single person. In order to improve algorithm robustness, multi-person postures need to be tested.
(2) The experiment is only at the stage of algorithm simulation. Nonetheless, our system hardware is completed, and testing work of transplanted algorithms will be done in the following stage.
(3) The final target is to deliver this wearable device to the elderly. More and more postures will be tested such as walking, sitting, sleeping, crawling, calling, falling down, and so on. Evaluating the weaknesses of the entire system and optimizing tasks need to be tested many times.

6. Conclusions

The paper provided an overview of a posture recognition method for elderly care. The technologies available can be divided into two categories: vision-based recognition and sensor-based recognition. To avoid invasion of privacy, sensor-based recognition is a better choice. We proposed a sensor-based scheme for posture recognition with the indoor positioning system and receiving tags. The positioning umbrella with seven arms can provide location services. Meanwhile, the receiving tags are pasted on the surface of special clothes to measure the distance from umbrella by UWB radio. In this solution, we carried out simulation experiments to verify the usability of the scheme. The LSE method and iEKF algorithm are introduced to estimate the positions of receiving tags. We also present posture recognition algorithms with structure vectors and posture angles which combined a couple of tags. Experimental results reveal that iEKF algorithm offers more accuracy than the LSE method, e.g., by calculating the coordinates of the head tag, the mean deviation of iEKF is 7.16 cm, lower than that of LSE (11.70 cm). In the improved extend Kalman filter, the influence of parameter N for window function has also been discussed, and the suggestion of reasonable range of N is given.

It has to be pointed out that there are also some disadvantages in our solution. We can achieve good posture recognition performance, but many tags are needed to keep working, and their energy is an important concern. In future work, we will transplant the algorithm to the processor, and focus on the further improvement of wearable technologies coupled with different kinds of postures, such as walking, sitting, sleeping, and so on. The test of the robustness and stability of the system also need to be carried out. We believe the prospect of applications for elderly care is vast.

Author Contributions: Conceptualization, X.H., W.F. and J.Z.; methodology, X.H. and W.F.; software, X.H. and J.J.; validation, Z.H. and J.J.; formal analysis, Z.H.; investigation, J.Z.; resources, J.Z.; writing—original draft preparation, X.H. and W.F.; writing—review and editing, J.Z.; visualization, X.H.; supervision, J.Z. and Z.H.; project administration, Z.H.

Funding: This research was jointly supported by the "Key Research and Development Projects of the Ministry of Science and Technology of China (2017YFD0701600)", the "National Natural Science Foundation of China (31401285)" and the "Natural Science Foundation of Department of Science and Technology of Anhui Province (1908085QF284)".

Acknowledgments: First and foremost, the authors would like to show deepest gratitude to Xuesen Li, a senior software engineer, who supported the team with the motion model in Unreal Engine 4.0. And then the authors would like to thank Dongsheng Yang for his excellent technical support and critically reviewing the manuscript.

References

1. Gong, S.; Wang, Y.; Zhang, M.; Wang, C. Design of Remote Elderly Health Monitoring System Based on MEMS Sensors. In Proceedings of the IEEE International Conference on Information and Automation (ICIA), Macau, China, 18–20 July 2017; pp. 494–498.

2. Aging Stats. Available online: https://agingstats.gov/ (accessed on 25 March 2019).

3. Yu, J.; Sun, J.; Li, W. 3D Human Pose Estimation Based on Multi-kernel Sparse Coding. *Acta Electron. Sin.* **2016**, *44*, 1899–1908.

4. Dai, Q.; Shi, X.; Qiao, J.Z.; Liu, F.; Zhang, D.Y. Articulated Human Pose Estimation with Occlusion Level. *J. Comput.-Aided Des. Comput. Graphics* **2017**, *29*, 279–289.

5. Kien, H.K.; Hung, N.K.; Chau, M.T.; Duyen, N.T.; Thanh, N.X. Single view image based-3D human pose reconstruction. In Proceedings of the 9th International Conference on Knowledge and Systems Engineering (KSE), Hue, Vietnam, 19–21 October 2017; pp. 118–123.

6. Hong, C.; Yu, J.; Tao, D.; Wang, M. Image-Based Three-Dimensional Human Pose Recovery by Multiview Locality-Sensitive Sparse Retrieval. *IEEE Trans. Ind. Electron.* **2015**, *62*, 3742–3751.

7. Tian, G.; Yin, J.; Han, X.; Yu, J. A Novel Human Activity Recognition Method Using Joint Points Information. *Robot* **2014**, *36*, 285–292.

8. Abbondanza, P.; Giancola, S.; Sala, R.; Tarabini, M. Accuracy of the Microsoft Kinect System in the Identification of the Body Posture. In Proceedings of the 6th International Conference on Wireless Mobile Communication and Healthcare, Milan, Italy, 14–16 November 2016; pp. 289–296.

9. Sombandith, V.; Walairacht, A.; Walairacht, S. Recognition of Lao Sentence Sign Language Using Kinect Sensor. In Proceedings of the 14th International Conference on Electrical Engineering/Electronics, Computer, Telecommunications and Information Technology, Phuket, Thailand, 27–30 June 2017; pp. 656–659.

10. Tripathy, S.R.; Chakravarty, K.; Sinha, A.; Chatterjee, D.; Saha, S.K. Constrained Kalman Filter for Improving Kinect Based Measurements. In Proceedings of the 2017 IEEE International Symposium on Circuits and Systems (ISCAS), Baltimore, MD, USA, 28–31 May 2017; pp. 1–4.

11. Gaglio, S.; Re, G.L.; Morana, M. Human Activity Recognition Process Using 3-D Posture Data. *IEEE Trans. Hum.-Mach. Syst.* **2015**, *45*, 586–597. [CrossRef]

12. Kinect for Windows. Available online: https://developer.microsoft.com/en-us/windows/kinect (accessed on 25 March 2019).

13. Pierleoni, P.; Belli, A.; Maurizi, L.; Palma, L.; Pernini, L.; Paniccia, M.; Valenti, S. A Wearable Fall Detector for Elderly People Based on AHRS and Barometric Sensor. *IEEE Sens. J.* **2016**, *16*, 6733–6744. [CrossRef]

14. Musalek, M. A wearable fall detector for elderly people. In Proceedings of the 28th DAAAM International Symposium, Zadar, Croatia, 8–11 November 2017; pp. 1015–1020.

15. Guo, G.; Chen, R.; Ye, F.; Chen, L.; Pan, Y.; Liu, M.; Cao, Z. A pose awareness solution for estimating pedestrian walking speed. *Remote Sens.* **2019**, *11*, 55. [CrossRef]

16. Wang, J.; Huang, Z.; Zhang, W.; Patil, A.; Patil, K.; Zhu, T.; Shiroma, E.J.; Schepps, M.A.; Harris, T.B. Wearable sensor based human posture recognition. In Proceedings of the IEEE International Conference on Big Data (Big Data), Washington, DC, USA, 5–8 December 2016; pp. 3432–3438.

17. Caroppo, A.; Leone, A.; Rescio, G.; Diraco, G.; Siciliano, P. Multi-sensor Platform for Detection of Anomalies in Human Sleep Patterns. In Proceedings of the 3rd National Conference Sensors, Rome, Italy, 23–25 February 2016; pp. 276–285.

18. Wang, Z.; Yang, Z.; Dong, T. A review of wearable technologies for elderly care that can accurately track indoor position, recognize physical activities and monitor vital signs in real time. *Sensors* **2017**, *17*, 341. [CrossRef]

19. Tian, X.; Li, W.; Yang, Y.; Zhang, Z.; Wang, X. Optimization of fingerprints reporting strategy for WLAN indoor localization. *IEEE Trans. Mob. Comput.* **2018**, *17*, 390–403. [CrossRef]

20. Zuo, Z.; Liu, L.; Zhang, L.; Fang, Y. Indoor positioning based on Bluetooth low-energy beacons adopting graph optimization. *Sensors* **2018**, *18*, 3736. [CrossRef]

21. Yasir, M.; Ho, S.-W.; Vellambi, B.N. Indoor position tracking using multiple optical receivers. *J. Lightware Technol.* **2016**, *34*, 1166–1176. [CrossRef]

22. Antoniazzi, F.; Paolini, G.; Roffia, L.; Masotti, D.; Costanzo, A.; Cinotti, T.S. A web of things approach for indoor position monitoring of elderly and impaired people. In Proceedings of the Conference of Open Innovation Association (FRUCT), Helsinki, Finland, 6–10 November 2017; pp. 51–56.

23. Dabove, P.; Di Pietra, V.; Piras, M.; Jabbar, A.A.; Kazim, S.A. Indoor positioning using ultra-wide band (UWB) technologies: positioning accuracies and sensors performances. In Proceedings of the IEEE/ION Position, Location and Navigation Symposium (PLANS 2018), Monterey, CA, USA, 23–26 April 2018; pp. 175–184.

24. Li, H.B.; Miura, R.; Nishikawa, H.; Kagawa, T.; Kojima, F. Proposals and implementation of high band IR-UWB for increasing propagation distance for indoor positioning. *IEICE Trans. Fundam. Electron. Commun. Comput. Sci.* **2018**, *E101A*, 185–194. [CrossRef]

25. Ridolfi, M.; Vandermeeren, S.; Defraye, J.; Steendam, H.; Gerlo, J.; De Clercq, D.; Hoebeke, J.; De Poorter, E. Experimental Evaluation of UWB Indoor Positioning for Sport Postures. *Sensors* **2018**, *18*, 168. [CrossRef] [PubMed]

26. Huang, J.; Yu, X.; Wang, Y.; Xiao, X. An integrated wireless wearable sensor system for posture recognition and indoor localization. *Sensors* **2016**, *16*, 1825. [CrossRef] [PubMed]

27. Xie, W.; Li, X.; Long, X. Underground operator monitoring based on ultra-wide band WSN. *Int. J. Online Eng.* **2018**, *14*, 219–229. [CrossRef]

28. Xia, B.; Lao, Z.; Zhang, R.; Tian, Y.; Chen, G.; Sun, Z.; Wang, W.; Sun, W.; Lai, Y.; Wang, M.; et al. Online parameter identification and state of charge estimation of lithium-ion batteries based on forgetting factor recursive least squares and nonlinear Kalman filter. *Energies* **2018**, *11*, 3. [CrossRef]

29. Xiao, M.; Zhang, Y.; Wang, Z.; Fu, H. An adaptive three-stage extended Kalman filter for nonlinear discrete-time system in presence of unknown inputs. *ISA Trans.* **2018**, *75*, 101–117. [CrossRef] [PubMed]

30. Huang, H.Y.; Chang, S.H. A skeleton-occluded repair method from Kinect. In Proceedings of the International Symposium on Computer, Consumer and Control, Taichung, Taiwan, 10–12 June 2014; pp. 264–267.

Article

A Novel Passive Indoor Localization Method by Fusion CSI Amplitude and Phase Information

Xiaochao Dang [1,2], Xiong Si [1], Zhanjun Hao [1,2,*] and Yaning Huang [1]

[1] College of Computer Science and Engineering, Northwest Normal University, Lanzhou 730070, China; dangxc@nwnu.edu.cn (X.D.); sx2016@nwnu.edu.cn (X.S.); bellanwnu@126.com (Y.H.)
[2] Gansu Province Internet of Things Engineering Research Center, Lanzhou 730070, China
* Correspondence: haozhj@nwnu.edu.cn

Received: 1 December 2018; Accepted: 15 February 2019; Published: 20 February 2019

Abstract: With the rapid development of wireless network technology, wireless passive indoor localization has become an increasingly important technique that is widely used in indoor location-based services. Channel state information (CSI) can provide more detailed and specific subcarrier information, which has gained the attention of researchers and has become an emphasis in indoor localization technology. However, existing research has generally adopted amplitude information for eigenvalue calculations. There are few research studies that have used phase information from CSI signals for localization purposes. To eliminate the signal interference existing in indoor environments, we present a passive human indoor localization method named FapFi, which fuses CSI amplitude and phase information to fully utilize richer signal characteristics to find location. In the offline stage, we filter out redundant values and outliers in the CSI amplitude information and then process the CSI phase information. A fusion method is utilized to store the processed amplitude and phase information as a fingerprint database. The experimental data from two typical laboratory and conference room environments were gathered and analyzed. The extensive experimental results demonstrate that the proposed algorithm is more efficient than other algorithms in data processing and achieves decimeter-level localization accuracy.

Keywords: indoor localization; channel state information; device-free passive; WiFi fingerprint; naive Bayes classification; feature fusion

---

1. Introduction

Location-based applications and services are concerned with people's daily lives, and thus have attracted increasing attention. According to different target environments, positioning services can be divided into two situations: outdoor localization and indoor localization. In the outdoor environment, traditional satellite positioning technologies, such as the Chinese BeiDou navigation satellite system (BDS), the global positioning system (GPS), and cellular-based station positioning technology provide highly-precise positioning services that can satisfy the needs of outdoor environment location services [1]. Indoor localization, as its name implies, is the positioning of the target person or object within an indoor environment, such as intrusion detection [2], security monitoring [3], and indoor navigation [4], among others [5]. Indoor localization requires timeliness, accuracy, and stability. However, in an indoor environment, the signal transmission will be limited by multipath interference, the shadow effect, power attenuation, transmission delay, etc., which lead to a poor performance of the positioning service [6]. Based on the above constraints, WiFi [7], Bluetooth [8], Radio Frequency Identification (RFID) [9], ultra-wideband (UWB) [10], and other wireless signal localization methods have been widely researched and applied [11].

Unfortunately, Bluetooth, RFID, UWB, and other wireless signal localization methods have their own shortcomings like requiring costly specific devices, being easily affected by external condition factors, or the equipment deployment being too complex. As WiFi technology matures and such devices are popularized, many indoor localization systems based on WiFi signals are being widely used to provide accurate and efficient location services because of the low cost, large range of signal transmission, and strong applicability [12–14]. Researchers have obtained received signal strength indicator (RSSI) signals from WiFi devices, analyzed the fluctuations caused by signal changes, established a signal propagation model, and mapped the RSSI signals to a distance, which serve as the basis for indoor localization [15,16]. Although the RSSI-based method has been greatly improved, the disadvantages of the coarse granularity and instability of the RSSI signal restrict the positioning effect [17].

With the use of orthogonal frequency division multiplexing (OFDM) systems and multiple-input multiple-output (MIMO) systems in the 802.11a/n protocol, channel state information (CSI) signals can be extracted from commercial WiFi equipment. In contrast to RSSI signals, which provide only amplitude information, CSI signals can provide both the subcarrier phase and amplitude information as well as better descriptions of the signal changes from the transmitter to the receiver than those provided by RSSI signals [18].

Device-free passive sensing is an emerging technology to sense humans or devices without attaching any additional device to them. The current research field of device-free wireless-based passive sensing is not limited to indoor localization, but also includes human behavior recognition, intrusion detection, which may go far in the future [19]. Generally, the passive indoor localization techniques do not require people to carry specific measuring devices which is more suitable for some special indoor places, such as elderly care, smart homes and security monitoring, etc. [20]. Traditional passive localization systems are mainly based on the coarse-grained RSSI signatures. Received signal strength indicator values usually vary greatly due to multipath even at the same position resulting in a limited localization accuracy [21]. WiFi is now accessible almost everywhere in daily life. Therefore, we can achieve passive indoor localization purposes by applying CSI which can also accomplish high accuracy position without the extra costs required for building infrastructure [22].

For instance, Xiao et al. [22,23] proposed the Fine-grained Indoor Fingerprinting System (FIFS) and the Pilot system. FIFS uses the diversity of the original CSI data in the time and frequency domains. It also utilizes a weighted average CSI value based on multiple antennas to improve the accuracy of indoor positioning. Pilot is the first proposal to leverage the temporal stability and frequency diversity characteristics of CSI for developing a CSI-based passive device-free indoor fingerprinting system. The authors of Reference [24] designed and implemented the novel dynamic multiple signal classification (Dynamic-MUSIC) method to detect the subtle reflection signals from the human body to identify the human target's angle for passive device-free localization. Chaapre et al. [25] designed a new method to generate position fingerprints called CSI-MIMO, which utilizes the phase and amplitude information of all subcarriers. However, the CSI-MIMO method produces only raw CSI data for each subcarrier without any processing and does not take advantage of multiple antennas, which could better reflect the uniqueness of the location. Zhefu Wu et al. [26] designed a naive Bayes classifier-based passive indoor localization system enhanced with confidence level information. Researchers have proposed PhaseFi, a phase fingerprinting system for indoor localization that involved designing a deep network with three hidden layers to train the calibrated phase data, and used weights to represent fingerprints [27]. However, whether all the CSI values collected from the Network Interface Card (NIC) contribute equally to the system accuracy has not yet been thoroughly studied. According to the authors of References [28,29], less significant features could be misleading and confuse the system, and these authors analyzed some factors causing instability in the CSI phase information and proposed linear transformation to remove the interference and extract phase features in order to realize localization.

So based on the above summary, we can conclude the several technical challenges of passive localization form these studies: (1) How to analyze and process raw CSI data to get stable or robust data features in indoor environments. (2) How to reveal the principle reasons for signal changes due to location difference, and mathematically model the relationship between the CSI fingerprint database and targets. (3) How to automatically cluster the different locations in large-scale fingerprint datasets and in short response times especially with high-accuracy requirements. However, these studies did not fully apply the fine-grained CSI amplitude and phase information, which made it impossible to achieve more accurate positioning. Therefore, it is urgent to solve the problem of passive indoor location with high precision.

To improve the indoor positioning accuracy and the overall effect, this paper presents FapFi, a passive indoor fingerprint system based on WiFi using the fusion amplitude and phase information of the CSI signal. About the acronym of FapFi, "F" represents the fusion method, "a" represents the amplitude of CSI, "p" represents the phase of CSI, and "Fi" represents the WiFi environment. This method determines the anomalous CSI amplitude values at the subcarrier level and the redundant values at the channel level. The amplitude information is filtered through the above process, and a linear transformation is then applied to extract the calibrated phase information. Finally, the fusion of the feature information is stored in a fingerprint database as a basis for indoor positioning. Compared with the techniques of the above-cited papers, the advantages of the approach described in this paper are as follows:

(1) We proposed to use a fine-grained physic layer information CSI for indoor localization and processed the CSI amplitude and phase data to obtain stable and robust fingerprint features while reducing the signal interference from environmental factors.
(2) We adopted a fusion method to extract the most contributing features from processed CSI data and constructed an efficient fingerprint database.
(3) FapFi applied Naïve Bayesian Classification which satisfies the real-time localization requirement for passive human indoor localization and high-precision positioning in two different environments.
(4) Regarding the performance of localization, we compared FapFi with other methods. We investigated the parameters that affect the performance of positioning accuracy. Experimental results demonstrated that the FapFi system is able to achieve high performance which outperforms a traditional CSI-based system in both environments.

This paper is organized as follows. In Section 2, we briefly introduce the background knowledge of indoor fingerprint localization and CSI. We introduce the amplitude and phase data cleansing methods and focus on the uniqueness and stability of this approach to localization in Section 3.1. Section 3.2 explains the structural design of the positioning system. Section 4 evaluates the selection of the experimental environment and the performance of the system by comparing the experimental results with those of other systems. Conclusions are presented in Section 5.

2. Preliminaries

In this section, we will present the background knowledge of indoor fingerprint localization and the FapFi system.

2.1. Fingerprint Localization

As shown in Figure 1, in an indoor environment, the wireless signal is obstructed by obstacles existing in the environment, and the signal is reflected and diffracted to form a multipath effect [30]. Different objects interfere differently with the transmission route, and the personnel are located in different locations; therefore, the signal characteristics are not the same. This difference can be invoked as a fingerprint feature. The process of fingerprint localization involves matching the signal characteristic of an unknown location with the existing information in a fingerprint database to match the best positioning result [31].

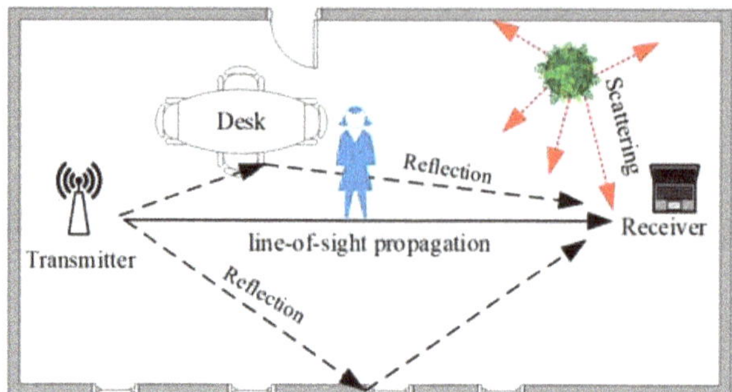

Figure 1. The path of signal transmission in an indoor environment.

The indoor fingerprint localization technology includes two stages: the offline stage and the online stage. The offline stage collects the position data of each reference point in the known area, processes the data to extract a feature, and establishes a relational database of the fingerprint feature of the reference point and the corresponding position coordinates. In the online positioning stage, the data are collected and analyzed in real time; the eigenvalues of the reference points are matched with the fingerprint database by using the matching algorithm, and the exact coordinate position results are thus obtained [32].

The traditional localization fingerprint feature is usually identified based on the RSSI signals, the signal-to-noise ratio (SNR), and other such parameters. However, the use of CSI has the advantages of temporal stability, frequency diversity, high stability, and providing a better reflection of the multipath effect in the environment. Furthermore, the applicability of CSI to the field of indoor localization is higher.

2.2. Channel State Information

The CSI signal contains more fine-grained physical layer (PHY) information during signal transmission and describes the signal characteristics, such as the amplitude and phase of each subcarrier wave in the channel. The CSI can better describe the communication link properties of the signal from the transmitter to the receiver, which can reflect the existing reflection, diffraction and other interference factors of an indoor environment. The CSI signal represents the combined effects of the channel status, such as scattering, fading, multipath interference, shadowing, and power decay with distance [33]. Currently, we can extract CSI signals in the frequency domain from the channel frequency response (CFR). Each CSI packet includes information such as the timestamp, RSSI, number of antennas, noise, and CSI.

The OFDM system divides the communication channel into several orthogonal subchannels with different frequencies. The received signal after a multipath channel transmission can be expressed as:

$$\vec{Y} = H\vec{X} + \vec{N} \tag{1}$$

where $\vec{Y}$ and $\vec{X}$ denote the signal vectors of the receiver and the transmitter, respectively, and H and $\vec{N}$ denote the channel information matrix and additive white Gaussian noise, respectively. The CSI of each subcarrier can be estimated from $\vec{X}$ and $\vec{Y}$ at the receiver, which is expressed as:

$$\hat{H} = \frac{\vec{Y}}{\vec{X}} \tag{2}$$

where $\hat{H}$ represents the CFR of each sub-channel. The CSI can be divided into different subcarrier groups according to the different hardware drivers at the receiver, and the CSI matrix can be expressed as:

$$H = [H_1, H_2, \cdots, H_N] \tag{3}$$

where N is the number of subcarrier divisions based on the hardware drivers of wireless network card. For example, $N = 56$ for a 20-MHz bandwidth channel, and $N = 114$ for a 40-MHz bandwidth channel. We used a 20-MHz bandwidth channel in this paper, thus the index of the subcarriers was 56. The CSI for each subcarrier is expressed as:

$$H_i = |H_i|e^{j\sin(\angle H_i)} \tag{4}$$

where $|H_i|$ and $\angle H_i$ are the amplitude and phase of the i-th subcarrier, respectively.

The CSI extracted from the experimental platform is an $m \times n \times k$ complex matrix, where m and n denote the numbers of antennas at the transmitter and the receiver, respectively, and k denotes the number of subcarriers.

The comparison between the CSI and RSSI signals is shown in Figure 2. In order to compare the effects of using CSI and RSSI data as data features, we randomly selected a test point in our environment and collected a set of CSI data packets from the Atheros 9380 network card. Meanwhile, we used Android mobile applications to collect RSSI data in the same environment in contrast to CSI data. The purpose of using mobile applications is to collect real-time RSSI data. The amplitude information of the CSI from three transmitter antennas and three receiver antennas can be analyzed using CSI data packets. It can be observed in the figure that the CSI signal changes more smoothly and steadily in the same environment and that the RSSI signal is prone to fluctuations. This comparison also verifies that the CSI signal has better spatial discrimination and stability over time, which is more suitable as a fingerprint for indoor localization.

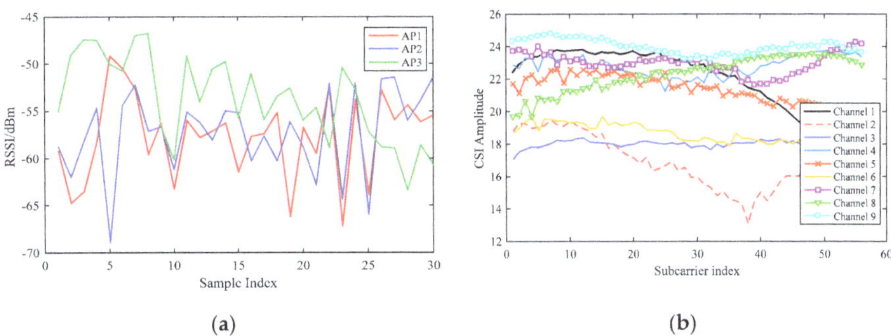

Figure 2. Comparison between (**a**) received signal strength indicator (RSSI) and (**b**) channel state information (CSI) signals.

2.3. Naive Bayesian Classification

The naive Bayes method is a classification algorithm based on Bayes' rule. The algorithm is easy to implement, and the overall complexity and usage of time and space is low [34]. In practical applications, this method has the advantages of dealing with multiple types of problems and making faster matches [35]. The principle is shown in Equation (5):

$$P(A|B) = \frac{P(B|A) \times P(A)}{P(B)} \tag{5}$$

where P is the probability and $P(A|B)$ is the conditional probability; A can be understood as a category, and B as a feature. The naive Bayesian classification is based on two basic assumptions: (1) the features are independent from each other; (2) each feature has the same probability distribution 26 [34].

The naive Bayesian classifier model is described as follows. Suppose the training set contains m classes $C = \{C_1, C_2, \cdots, C_m\}$ and n conditional attributes $X = \{X_1, X_2, \cdots, X_n\}$. Assuming that all of the conditional attributes X are children of a class variable C, if $P(C_i|X) > P(C_j|X)$, and only if $(1 \leq i, j \leq m, i \neq j)$ holds, then assign a given sample to be classified $X = \{x_1, x_2, \cdots, x_n\}$ to class $C_i (1 \leq i \leq m)$. According to Equation (5), the posterior probability of class C_i is:

$$P(C_i|X) = \frac{P(C_i)P(X|C_i)}{P(X)} \tag{6}$$

In the classification problem of this paper, C indicates the set of location points to be classified and X represents the set of CSI fingerprint feature data at a location point. The specific application will be discussed in Section 3.2.

3. FapFi System Design

The CSI-based fingerprinting mainly needs to meet two requirements: (1) the signal should be fixed as stable as possible when the CSI signal is at a certain point; and (2) the CSI signals collected at different points should be easily distinguished to distinguish different locations. As shown in Figure 3, the red and blue CSI signal amplitudes were collected at different points and can reflect the stability and distinguishability of the CSI. Furthermore, the figure also reflects that due to several factors such as strong multipath effect, the position of the furniture and so on in the environmental interference, there is considerable noise in the original CSI signal. The FapFi method fuses processed CSI amplitude and phase data to obtain stable and robust fingerprint features. The most important thing is FapFi satisfied timeliness, accuracy, and stability requirement of passive localization. The following content will introduce our main idea of designing FapFi method.

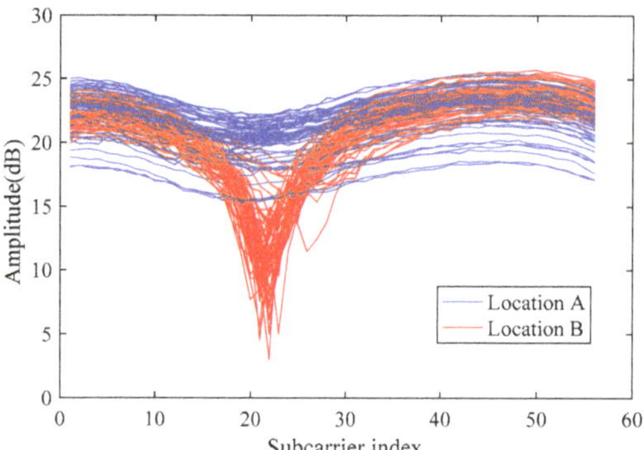

Figure 3. CSI amplitude.

To reduce the CSI signal interference caused by the indoor environment, we need to solve two key factors including data preprocessing and feature extraction, thereby we try to ensure that the experimenter is absolutely static and the test environment is stable when collecting data in the offline and online stage, which can reduce other interference factors such as changes in furniture position. Secondly, we choose a relatively empty environment to collect a large number of experimental data for analysis and then we utilize the phase characteristics of the CSI to make up for the shortcomings of the amplitude. In this paper, we separately processed the amplitude and phase information, and the processed features were then fused to improve the positioning accuracy.

3.1. Data Sanitization

In this section, we describe the data cleansing step and utilize the fusion feature to build a unique, robust fingerprint.

3.1.1. Amplitude Sanitization

During the process of indoor localization, the tester was required to stand still, and changes in the signal due to slight activity of the human body will thus be concentrated at frequencies lower than any abnormal frequencies [36]. Based on this analysis, the filter of the amplitude can effectively remove the irrelevant signal frequencies caused by non-human activities, thereby eliminating the noise interference caused by the amplitude of the subcarriers while preserving the effective amplitude data.

We had a researcher remain standing at the test point and continuously collected 100 CSI packets, taking the amplitude of one of the links. The results are presented in Figure 4a,b. The redundant values and outliers of the CSI signals are indicated by the arrows in the figure. To judge the abnormal values present in the collected CSI signal amplitudes, we chose to calculate the standard deviation for the collected data packet at the subcarrier level. The standard deviation describes the degree to which the amplitude deviates from the mean value during the data acquisition stage, so it can be used to judge the outliers of the amplitude. Regarding the redundant values, we considered the relationship of the outlier data packets in the filtering process to ensure the correlation between the data in the channel level, which means that during the entire communication process we processed the redundant values. The specific steps in filtering the CSI amplitude characteristics are shown below.

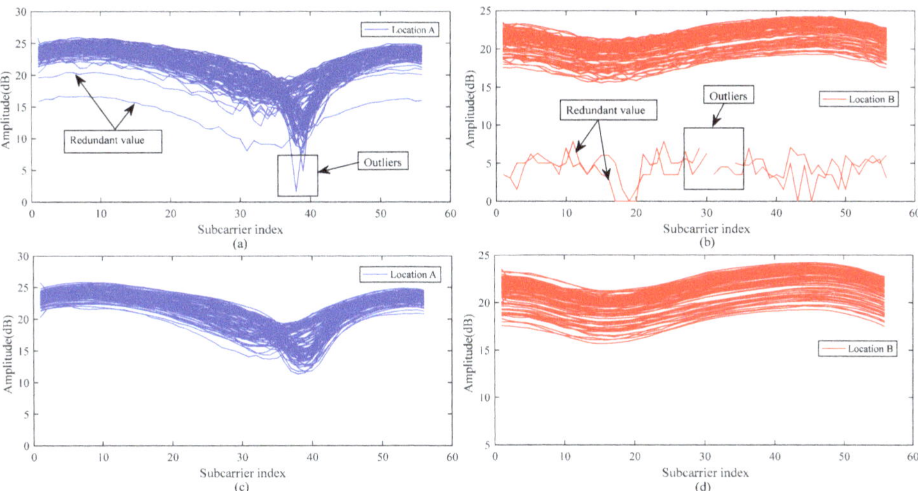

Figure 4. (**a**,**c**) unprocessed CSI amplitudes; (**b**,**d**) processed CSI amplitudes.

Step 1: Calculate the mean value $\overline{Am^i}$ of the i-th subcarrier of k-th data packet according to Equation (7):

$$\overline{Am^i} = \frac{1}{N} \sum_{k=1}^{N} Am_k^i \tag{7}$$

where N is the number of samples, and $i \in [1, 56]$ is the subcarrier index.

Step 2: Calculate the standard deviation of the i-th subcarrier from Equation (8):

$$\sigma_i = \sqrt{\frac{1}{N} \sum_{k=1}^{N} \left(Am_k^i - \overline{Am^i} \right)^2} \tag{8}$$

i is the index of the subcarriers, so we can get the $V = [\sigma_1, \sigma_2, \cdots, \sigma_{55}, \sigma_{56}]$ which is a variance matrix of the 56 subcarriers.

Step 3: Assuming that the data packet to be filtered is k, the CSI amplitude values are $|Am|^i_{k-1}$ and $|Am|^i_{k+1}$ for each adjacent data packet $k-1$ and $k+1$, respectively. According to Equation (8), the filtered amplitude $|Am|^i_{filter}$ is calculated by averaging the three amplitude data values.

$$|Am|^i_{filter} = \frac{1}{3}(|Am|^i_{k-1} + |Am|^i_k + |Am|^i_{k+1}) \tag{9}$$

Step 4: For the processed amplitude Am_{filter}, the covariance matrix $Cov(Am_{filter}, Am)$ of Am_{filter} and $\overline{Am}$ is calculated. If the variation trends of the two variables are consistent, the covariance is positive. If the two variables change in opposite directions, the covariance between the two variables is negative, which is considered redundant and removed from the packet.

As shown in Figure 4a,b, after the above steps determination, we can observe that the redundant values are usually deviated from normal values which means that the covariance between redundant values and normal values are negative. Outliers are usually caused by problems such as data packet loss or data peak anomaly.

After processing the CSI amplitude values shown in Figure 4c,d below, the filtered CSI amplitude values are smoother, with the redundancy caused by various factors effectively removed and the abnormal values caused by environmental factors filtered out. In the course of the experiment, we adopted a large number of CSI packets, so the problem that outlier filtering processing over-sensitive and inaccurate data can be avoided.

3.1.2. Phase Sanitization

The phase information is seldom used for CSI-based indoor localization schemes mainly because the hardware is not good enough to measure the true phase [37]. Therefore, researchers generally use the amplitude value of the CSI to locate an object, and the phase is rarely used in indoor localization. The phase measurement $\angle \hat{Ph_i}$ of the i-th subcarrier can be expressed by Equation (10):

$$\angle \hat{Ph_i} = \angle Ph_i + 2\pi \frac{k_i}{N} t + \beta + Z \tag{10}$$

where $\angle Ph_i$ is the true phase value, t is the timing offset between the transmitter and receiver, β is the phase offset caused by the carrier frequency offset, and Z is the measurement noise, while k_i stands for the i-th subcarrier index. In the Atheros platform $k \in (1, 56)$, N is the number of fast Fourier transform (FFT) samples, and N is 64 in the IEEE 802.11 a/g/n protocol. Due to the above factors t, β, and Z, ordinary WiFi NICs are unable to obtain the true phase values.

To mitigate the impact of random noises and extract the available phase information, we measure changes of CSI signal phase in an indoor environment and perform a linear transformation on the raw phase data, as recommended in References [28,36,38,39] et al. The main idea is to apply a linear transformation to the original phase value and eliminate the interference factors t and β by considering the phase across the entire frequency band.

First, the two formulas for slope a and intercept b are defined as follows:

$$a = \frac{\angle \hat{Ph_i} - \angle \hat{Ph_1}}{k_i - k_1} = \frac{\angle Ph_i - \angle Ph_1}{k_i - k_1} - \frac{2\pi}{N}\Delta t \tag{11}$$

$$b = \frac{1}{n}\sum_{j=1}^{n} \angle Ph_j - \frac{2\pi \Delta t}{nN}\sum_{j=1}^{n} k_j + \beta \tag{12}$$

Equations (11) and (12) are based on the assumption of the linear transformation method. In Equation (11), Δt is the corresponding difference of timing offset. According to the IEEE 802.11n specification, the subcarrier frequency is symmetric, which indicates $\sum_{j=1}^{n} k_j = 0$, so the error term Δt

can be further removed and b can be expressed as $b = \frac{1}{n}\sum_{j=1}^{n} \angle Ph_j + \beta$, which neglects the influence of the measurement noise Z. β is the equivalent timing offset of the receiver caused by the device, which in fact cannot be eliminated but can only reduce its impact. By subtracting the linear term $ak_i + b$ from the original phase $\angle \hat{Ph}_i$, part of the random phase shifts can be removed. We can then obtain the transformed phase denoted by $\angle \tilde{Ph}_i$:

$$\angle \tilde{Ph}_i = \angle \hat{Ph}_i - ak_i - b = \angle Ph_i - \frac{\angle Ph_n - \angle Ph_1}{k_n - k_1}k_i - \frac{1}{n}\sum_{j=1}^{n} \angle Ph_i \tag{13}$$

Figure 5a–c depicts the raw phase features of channel 1, channel 5, and channel 9, respectively, with six data packets selected for each channel to illustrate the effect that CSI phase information is unstable in the environment. The CSI phase was also affected by the sampling frequency offset and the carrier frequency offset; while there has been some research conducted on this, we did not discuss these two factors here, we only calibrated the phase information through the handy linear transform method to apply phase features in the fingerprint database. After linear processing, the instability of the phase was effectively reduced and the phase features could be converted into ordered and analyzable data. Figure 5d–f shows the processed phase of the six CSI data packets; the raw phase had a large fluctuation range and the processed phase had a smoother, small range especially where the original phase fluctuated. This indicates that the proposed linear transformation can remove the phase offset. Overall, the calibrated phase is stable enough for indoor localization requirements and solved the problem of data instability, which is the basis for extracting the high-robustness and high-efficiency fingerprint feature values by fusion of the amplitude and phase features.

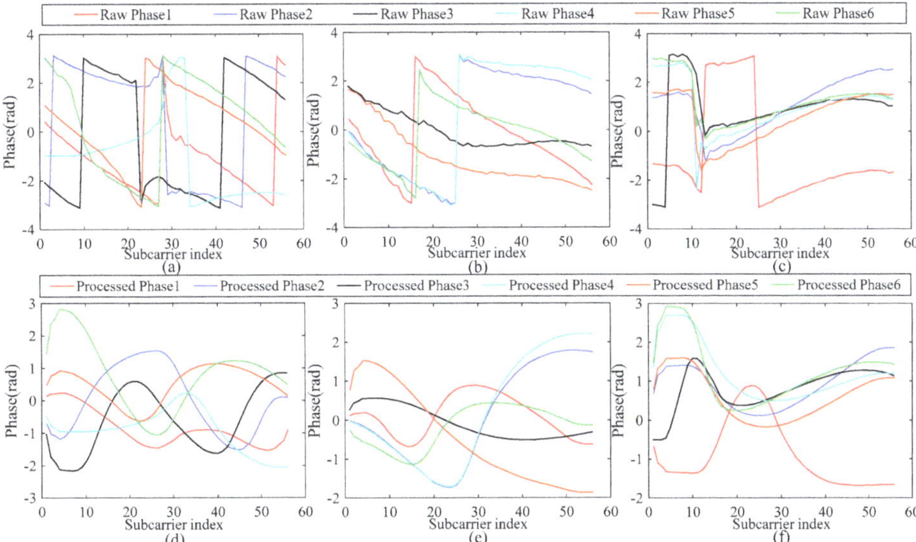

Figure 5. (**a–c**) Original phase for the 6 data packets in channels 1, 5, and 9; (**d–f**) corresponding phase processing result for a set data packets in different channels.

3.2. System Architecture

The FapFi system architecture is shown in Figure 6. In the offline phase, P reference points are selected in the target area L; the position information of each reference point is known, and the CSI values of all the reference points at times Q are collected to form the original position fingerprint

$$F:F = \begin{vmatrix} csi_{11} & csi_{12} & \cdots & csi_{1Q} \\ csi_{21} & csi_{22} & \cdots & csi_{2Q} \\ \vdots & \vdots & \vdots & \vdots \\ csi_{P1} & csi_{P2} & \cdots & csi_{PQ} \end{vmatrix}$$. Equation (4) shows that $csi_{PQ} = H_{PQ}$; that is, the amplitude and phase information of each point are obtained.

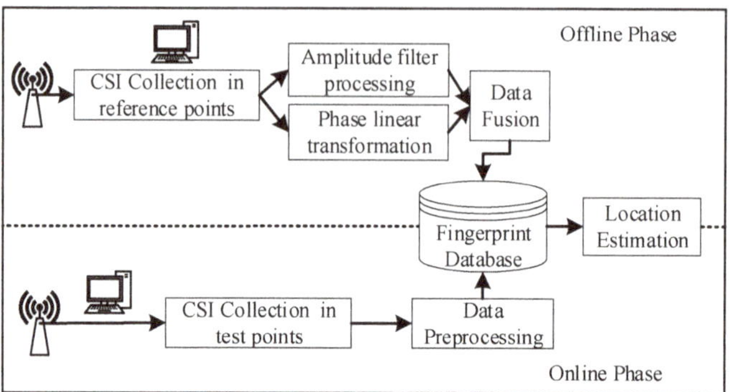

Figure 6. Positioning system architecture.

Assuming that the processed amplitude is $|Am|^{new}$ and the phase is $\angle \widetilde{Ph}$, a linear weighted fusion is performed on $|Am|^{new}$ and $\angle \widetilde{Ph}$ to obtain a new feature csi':

$$csi' = w_1 |Am|^{new} + w_2 \angle \widetilde{Ph} \tag{14}$$

Equation (14) can be used to calculate the percentage of the amplitude or phase in the new fingerprint database, and it represents the proportion of the amplitude and phase information of a fingerprint feature taken at a test point after it has been processed. It is convenient and simple to calculate using the linear weighted fusion method. Where w_1 and w_2 are the feature fusion weights, which satisfy the constraints of Equation (15), and csi' is the fused feature.

$$\begin{cases} w_1 + w_2 = 1 \\ 0 < w_1 < 1 \\ 0 < w_2 < 1 \end{cases} \tag{15}$$

The fingerprint feature of a test point is only composed of CSI amplitude and phase, and its condition is constrained by Equation (15). In advance, set $w_1 = w_2 = 0.5$ and adjust the weight of the amplitude and phase in the fingerprint database according to the changes in the environment and the positioning accuracy; that is, dynamically adjust the w_1 and w_2 assignments, the details will be discussed in Section 4.2.3.

The original amplitude and phase information of P reference points are subjected in Q times to the above processing steps to form a new signal feature csi_{PQ}', thereby developing the fingerprint database F' in the offline phase and forming the mapping relationship between the points in the spaces L and F. At present, there is a ubiquitous limitation of fingerprint positioning. The fingerprint database is only available for the environment and targets we have built. If the environment and target are changed, the database needs to rebuild.

In the online phase, the signal features of an unknown position $l_i(x_i, y_i)$ are collected, and the amplitude and phase information of the point is obtained after the data processing. The naive Bayesian classification algorithm is matched with the fingerprint database F' online to output the best result that estimates the location of the test point l_i.

For prior probabilities of random locations $l_i \in L$ in a space to be the same and known, $P(l_i|x)$ is equivalent to calculating the maximum posterior probability of $P(x|l_i)$, where x denotes the term to be classified in naive Bayesian algorithm:

$$P(l_i|x) = \frac{P(l_i)P(x|l_i)}{P(x)} \tag{16}$$

Assuming that $P(l_i)$ and $P(x)$ are known, the probability estimation obeys the Gaussian distribution, $P(x|l_i) \sim N(\delta, \theta)$, δ and θ are the mean deviation $mean(i)$ and mean square error $std(i)$, respectively. The maximum probability value $P(x|l_i)$ of location l_i is solved by Equation (17), and the category with the largest posterior probability is taken as the matching result of the unknown point l_i.

$$l_i \leftarrow \mathrm{argmax}\, P(l_i|x) = \mathrm{argmax} \frac{P(L_i)P(x|L_i)}{P(x)} \tag{17}$$

The predictive ability of the Bayesian classification algorithm is related to the completeness of the training samples. After adopting the correlation processing method of this algorithm, the training samples are more representative, which can lead to more accurate location results.

4. Experiment Validation

In this section, we show the detailed results obtained in the experimental environment and the data collection method of our proposed system. Afterward, we will evaluate the performance of our proposed system in various scenarios and compare the resulting location errors in different environments with several benchmark schemes.

4.1. Experimental Setup

In FapFi system, two desktop computers with Atheros 9380 network cards were employed in the experimental environment. The Ubuntu10.04LTS operating system (Canonical, London, England), which is driven by a custom kernel and modified wireless network card firmware, was installed in two desktop computers. One PC was equipped with an Intel Core i3-4150 CPU (Intel, Santa Clara, CA, USA) that functioned as a transmitter, and the other PC worked as a receiver. The Atheros NIC programs used the Atheros-CSI-Tool, which is an open source driver developed by Xie et al. [40]. After the above procedures, the Atheros 9380 wireless network card can extract the CSI signals between the receiver and the transmitter. As shown in Figure 7, the Atheros 9380 NIC we used had three antennas in the experiment.

Our techniques were tested in two different indoor sceneries. The first testing scenery was a relatively empty laboratory which has been commonly used in many previous studies. As shown in Figure 8a, there were enormous line of sight (LOS) receptions which suffer less from multipath effects to validate the effectiveness of the system. In the 9 m × 6 m area, 25 deployment square areas were deployed. Each square area was 0.8 m × 0.8 m, the center of the square was the corresponding position coordinate of the reference point in the process of building the fingerprint database, and the receiver antennas was 4.5 m apart from the transmitter antennas and the antenna height was 1.2 m.

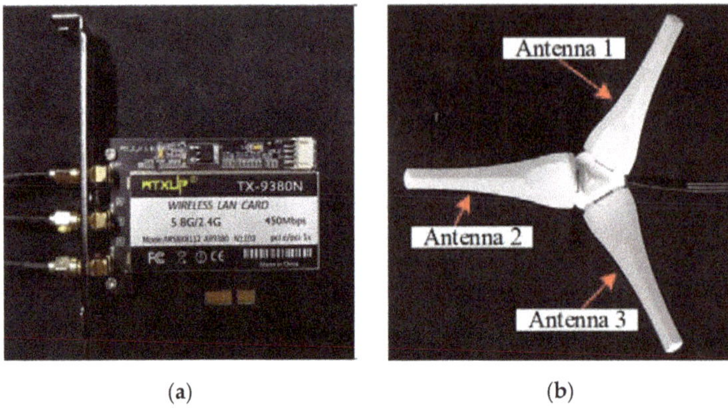

(a)   (b)

Figure 7. Experimental equipment (**a**) Atheros 9380 network cards; (**b**) NIC antennas

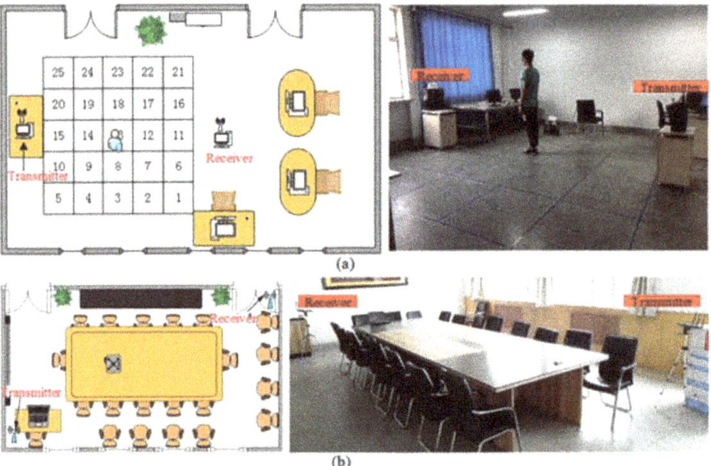

Figure 8. Experimental scenarios: (**a**) floor plan in laboratory; (**b**) floor plan in meeting room and experimental equipment.

The approaches proposed in previous studies were primarily tested in the first environment which contained less noise and interference. The second testing scenery was a meeting room with metal tables, chairs, and desktop computers. Hence, this environment had a relatively large number of multipaths in more extreme environments to further validate our method with previous work. Figure 8b shows the environment of the second testing location, which was quite crowded, and most of the LOS paths were blocked. The experimental equipment configuration was as described above. We arbitrarily chose test points within the 8 m × 6 m area. The receiver antennas were 6.5 m apart from the transmitter antennas, and the antenna height was 1.2 m. In both testing cases, the direction and position of the experimenter's stance remained unchanged during the data collection and estimation.

4.2. Experimental Analysis

In this section, we discuss various parameters and evaluate their impact on the performance of our system, such as the selected fingerprint features, the number of antennas deployed, and test samples. We illustrate the results in the rest of paper.

The localization effect of the algorithm can be measured by two indexes: location accuracy and average error distance.

Location Accuracy: The ratio of the correct location prediction category to the total number of tests.

Average Error Distance: Assuming that the number of tests is N, the i-th estimated position coordinates $\hat{L}(\hat{x}_i, \hat{y}_i)$, the actual position coordinates $L(x_i, y_i)$, the error distance can be obtained by Euclidean distance between $\hat{L}$ and L, and the average error distance D_{error} can be expressed by Equation (18):

$$D_{error} = \frac{1}{N} \sum_{i=1}^{N} \sqrt{(\hat{x}_i - x_i)^2 + (\hat{y}_i - y_i)^2} \tag{18}$$

4.2.1. Impact of Selected Fingerprint Features

In the experiment, the tester (with a height of 1.83 m) stood in the test area to collect the test data for different test points. We separately analyzed the results obtained using the amplitude and phase fusion, the processed amplitude, the raw amplitude, and RSSI fingerprinting. Because the original CSI phase data were not useful for the location, the original phase data were not analyzed in this experiment. The location error for the cumulative distribution function of the fingerprint database constructed with the various feature data is shown in Figure 9. As seen from Figure 9a, using CSI signals as the fingerprint feature for indoor localization is better than using RSSI signals in the laboratory environment. A positioning accuracy of 1.8–2.5 m could be achieved with RSSI signals, and using the processed amplitude as the eigenvalue resulted in sub-meter-level positioning accuracy. Through the fusion of the processed amplitude and phase data, the positioning error for 90% of the test points could be reduced to within 1 m, with a probability that 54.6% of the test points were within 0.5 m. The overall positioning accuracy was affected by the complex environment in the meeting room; the tables and other objects obstructed most LOS paths and magnified the multipath effect. In the more complex environment, a positioning accuracy of 2–3 m could be achieved with RSSI signals, and using the amplitude and phase fusion method resulted in a 1-m distance error for over 56% of the test points. The overall performance of this fusion method was better than any other method, which greatly enhanced the CSI-based indoor localization accuracy and verified the validity of the amplitude and phase information processing proposed in this paper. Figure 9 also illustrates that the unprocessed CSI values will affect the accuracy of the decline, and the amplitude and phase data cleansing methods proposed in this paper are effective for improving positioning performance.

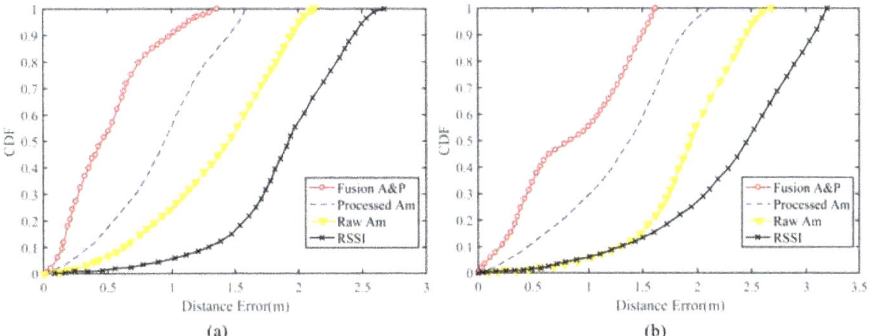

Figure 9. **Cumulative Distribution Function** (CDF) of localization error of different fingerprint features in different environment (**a**) in the laboratory and (**b**) in the meeting room.

4.2.2. Impact of Antennas and the Number of Packets

With the indoor location algorithm based on WiFi, the numbers of deployed antennas, training samples, and testing samples are key factors affecting the positioning accuracy when using the fingerprint location method. The numbers of transmitting and receiving antennas determine the number of channels, and the optimal combination of antenna numbers can achieve highly precise positioning accuracy. To address this issue, in this experiment, we deployed the receiver and transmitter shown in Figure 7, and we set $m = 3, n = 1, m = 3, n = 2$, and $m = 3, n = 3$, where m and n denote the numbers of antennas at the transmitter and the receiver, respectively. So, we can get a combination of 3, 6, and 9 channels, and it is worth noting that the concept of channels are communication links which act as signal transmission paths between transmitters and receivers. Then we evaluated six combinations: 1000 training data packets/200 testing data packets, 500 training data packets/200 testing data packets, 500 training data packets/100 testing data packets, 250 training data packets/100 testing data packets, 100 training data packets/100 testing data packets, and 100 training data packets/50 testing data packets, each performed with 3, 6, and 9 channels. A total of six test samples were experimentally tested to analyze the corresponding average positioning error in a laboratory scenario.

Figure 10 shows that, without considering the impact of data packets, the average distance error of positioning was the highest among all the sample combinations when only three channels were used. As the number of antennas increased, the positioning accuracy gradually increased.

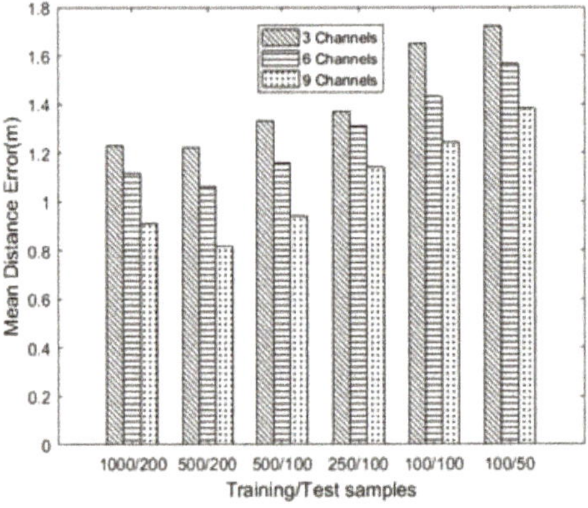

Figure 10. Positioning accuracy of different antennas and packet numbers.

Next, we analyzed the influence of the number of training samples and the number of test samples on the positioning accuracy. The average positioning error for the 1000/200 data and the 500/200 data combinations was relatively small. However, for the 1000/200 combination, the main reason for the slight performance degradation compared to the 500/200 combination was that the 1000/200 combination required a longer time to train the classification model; in addition, there was overfitting when the training dataset needed to extract and analyze more amplitude and phase information in data processing. For the 250/100, 100/100, and 100/50 data sets, the localization effect was not ideal when the data samples were relatively few. Based on the experimental results, 500 training packets/200 test packets with nine channels were used for subsequent experiments.

4.2.3. Impact of Feature Fusion Weights

In Section 3, we used a linear weighted fusion in constructing the fingerprint database, with the feature fusion weights w_1 and w_2 representing the proportion of the amplitude and phase information, respectively, in the fingerprint database. After data fusion, the feature data not only effectively reduced the data dimensions, but also makes full use of CSI fine-grained data, and can more reasonably allocate the proportion of amplitude and phase features in the fingerprint database.

In order to eliminate redundant information and generate more distinctive fingerprint features, we dynamically adjusted the weights according to the constraints of weights. In the initial stage, we set $w_1 = w_2 = 0.5$ and tested it 10 times for debugging. We can calculate the maximum positioning error $\max D_{error}$ and the minimum positioning error $\min D_{error}$ by Equation (18), and then calibrate next weights according to the latest positioning error. The weight value of amplitude w_1 is calculated by Equation (19) and the weight value of phase w_2 was obtained. In order to ensure the accuracy of positioning, we specified that the values of w_1 and w_2 must not be less than 0.1 or greater than 0.9. Also, if $D_{error} > \mathrm{max}D_{error}$ or $D_{error} < \mathrm{min}D_{error}$, the value of $\mathrm{max}D_{error}$ or $\mathrm{min}D_{error}$ was updated immediately.

$$w_1 = \frac{D_{error} - \mathrm{min}D_{error}}{\mathrm{max}D_{error} - \mathrm{min}D_{error}} \tag{19}$$

To analyze the influence of the feature fusion weight values w_1 and w_2 on the localization result, we used nine groups of control data to carry out the experiments, the case of $w_1 = 1$ or $w_2 = 1$ was already discussed in Section 4.2.1. We ran 100 tests in the laboratory environment and the meeting room environment to calculate the location accuracy under different conditions. The results are shown in Figure 11. When using the dynamic adjustment scheme, the accuracy of positioning in laboratory and meeting room environments reached 81% and 75%, respectively. It can be observed that the positioning accuracy of the dynamic adjustment weight method in different environments was higher than that setting the fixed weight. Although the dynamic adjustment weight value method was not obvious for the improvement of the positioning accuracy, it provided flexibility for the improvement of the positioning error. The results of these experiments indicate that the phase information can compensate for shortcomings when the amplitude information performs poorly at the more accurate positioning distance. The appropriate adjustment of the weights of the amplitude and phase in the fingerprint database can reasonably utilize the fine-grained data features of CSI and improve the positioning accuracy.

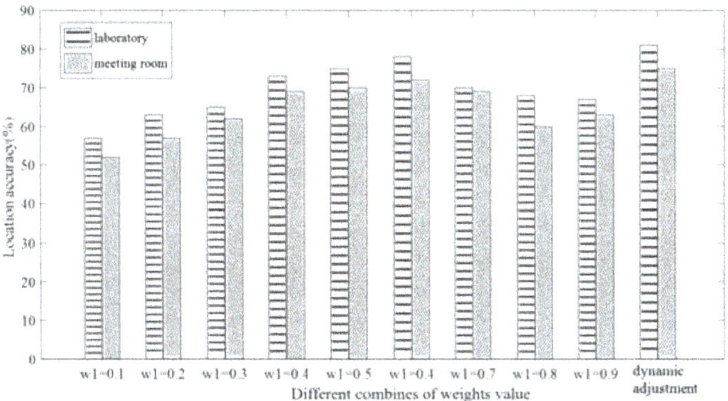

Figure 11. Influence of feature fusion weights on positioning accuracy in different environments.

4.2.4. Impact of the Number of Reference Points

In the fingerprint localization algorithm, the number of reference points used in the fingerprint space in the offline stage is also an important parameter that affects the positioning performance. More reference points result in a better positioning effect. In this paper, to analyze the impact of the number of reference points on the localization algorithm, according to the planar design in Figure 8a, we selected 25, 50, 75, and 100 reference points as the parameter variables to analyze the effect on the results and the execution time of the algorithm in a laboratory scenario.

As shown in Figure 12, in the selected experimental region, when the number of reference points was selected in the range of 75 to 100, the positioning accuracy was approximately 0.75 m, and the execution time of the algorithm was approximately 2 s. An analysis based on these two aspects yields an optimal positioning result. In terms of the overall analysis, the number of reference points was positively correlated with the positioning accuracy. As the number of reference points increases, the positioning error gradually decreases, while the time taken for offline fingerprint database acquisition was not considered. Compared with the online phase algorithm, increasing the number of reference points will increase the algorithm execution time.

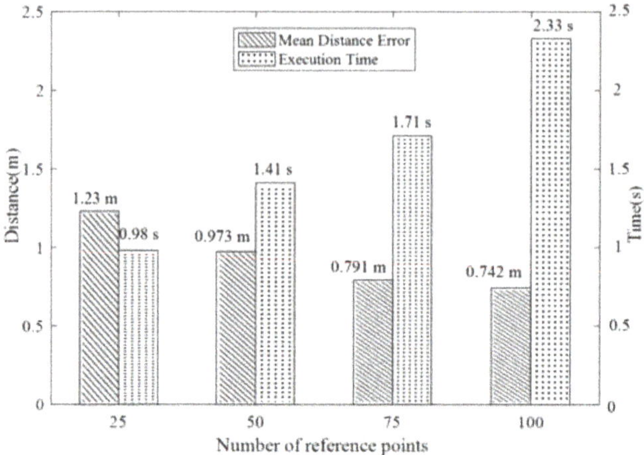

Figure 12. Influence of reference points on positioning accuracy.

4.2.5. Overall Performance

To compare the effects of different localization algorithms, we compared our proposed localization method with the PhaseFi, CSI-MIMO, and FIFS systems. We collected 500 training packets/200 test packets from test locations with nine channels and randomly selected test locations in the environments. We validated our technique under two environmental conditions.

As shown in Figure 13a, in the proposed system, 58.8% of the test position errors were controlled within 0.5 m, and 91.1% of the errors were below 1 m, while PhaseFi ensured that approximately 21% of the test positions had errors below 0.5 m and 87% were below 1 m. The overall performance of the CSI-MIMO and FIFS systems in the test environment was not good, mainly because FIFS only uses the diversity of the original data in the time domain and the frequency domain as the fingerprint database and because CSI-MIMO uses the multi-antenna mode to collect the CSI data. Neither of these systems further processed the data, thus leading to an overall positioning accuracy of greater than 1 m. For PhaseFi, which uses neural network algorithms, the advantages were reflected by positioning accuracies in the range of 0.7–1 m. FapFi had a higher positioning efficiency than PhaseFi within 0.3–0.7 m because FapFi adopts the fusion method to utilize both the amplitude and phase features of the CSI.

The meeting room environment contrasts with the laboratory environment. As shown in Figure 13b, although multipath propagation degrades the accuracy of the localization, our proposed method is robust enough to maintain accuracy in the meeting room. The positioning accuracy of FapFi within 1 m was still very good, whereas with PhaseFi, only 42% of the test points had an estimation error under 1 m, while for CSI-MIMO and FIFS, the values were 27.5% and 19%, respectively. Our proposed system performed better than the other algorithms within the range of 1–1.5 m. It can be concluded from the experiment that the use of phase features can compensate for the shortcomings of using only the amplitude feature for the fingerprint database in the traditional method and can achieve a better positioning effect. The fundamental reason for this is because of the way in which we processed the data, which was very effective.

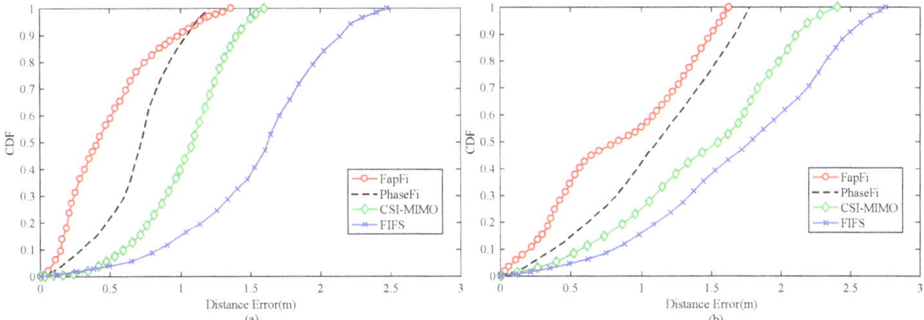

Figure 13. CDF of distance errors using different localization methods in different environments: (**a**) in the laboratory and (**b**) in the meeting room.

We also analyzed the average execution time of different systems for a further comparison. The execution times of different systems included reference point CSI data acquisition time, data processing time, and position estimation time. Reference point CSI data acquisition time is offline stage data acquisition time, and data processing time is calculated by the time-consuming in the data sanitization stage and position estimation time is the response time of the different systems in the online stage. Figure 14 shows the execution time required by various algorithms for a single test point. In the CSI collection procedure, FapFi, PhaseFi, and FIFS need the same amount of time to collect the data. The CSI-MIMO uses multiple antennas for the training dataset, and thus requires more collection time. In the processing stage, PhaseFi takes 2.1 s longer than the other algorithms, and only the FIFS method takes less time to build the fingerprint database because FIFS uses a weighted average value of different antennas and the coherence bandwidth to reduce the complexity of the algorithm. During the online phase, the estimation procedure of the proposed system is 1.41 s faster than the other systems. The main reason for this is that we used the naive Bayesian algorithm, and the advantage of the naive Bayesian algorithm is that it enables a viable and effective classification of large data sets in a relatively short time. The overall results show that the total execution time of FapFi is 4.11 s and that of PhaseFi, CSI-MIMO, and FIFS is 4.91 s, 5.21 s, and 3.43 s, respectively. Due to the superiority of the FIFS system in the preprocessing stage, FIFS takes less execution time, but the FIFS system positioning accuracy is not high. In contrast, FapFi can achieve greater positioning accuracy in a shorter amount of time than the other algorithms in terms of location accuracy and execution time.

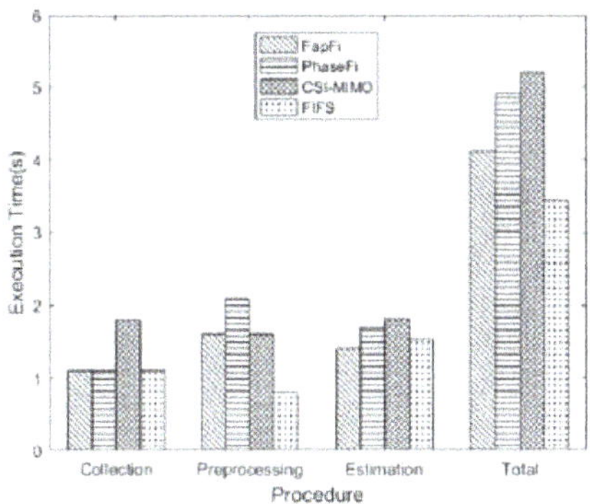

Figure 14. Execution time of different localization schemes.

5. Conclusions

We considered the multipath effect and time-variance of CSI signals in indoor environments. In this paper, we proposed FapFi, a passive indoor localization system leveraging the CSI amplitude and phase features to yield the fingerprint. FapFi takes advantage of the physical layer of the CSI in the widely used, off-the-shelf WiFi infrastructure and aims to achieve a high-precision positioning effect. To eliminate the interference caused by signals in the indoor environment, we filtered the CSI amplitude information and linearly converted the phase information. Furthermore, the performance of our system can be further enhanced by utilizing the CSI amplitude and phase information after fusion processing. We validated the performance of our system in both a laboratory and a meeting room. According to the experimental results, the average positioning error was approximately 0.5 m in the laboratory and within 1.2 m in the meeting room. Then, compared with the CSI-based localization method, our presented fingerprint had higher accuracy, which verifies that the proposed fusion of the CSI amplitude and phase data can effectively improve the accuracy of indoor positioning.

In future work, the flexibility of the system is expected to improve step by step, and the timeliness of the fingerprint database will be further improved. More importantly, the experimental deployment and testing are very complex and time-consuming as of now. With the release of the Atheros CSI Tool OpenWRT version, our future research work will find a better way to solve the site survey or fingerprint databases construction, such as a crowdsourcing approach or other machine learning algorithms [41–44].

Author Contributions: X.D. contributed towards the algorithms and the analysis. As the supervisor of X.S., he proofread the paper several times and provided guidance throughout the whole preparation of the manuscript. X.S. contributed towards the algorithms, the analysis, and the simulations and wrote the paper. Z.H. and Y.H. revised the equations, helped in writing the introduction and the related works, and critically revised the paper. All authors read and approved the final manuscript.

Funding: This research was funded by the National Natural Science Foundation of China under Grant No. 61762079, and No. 61662070, Key Science and Technology Support Program of Gansu Province under Grant No. 1604FKCA097 and No. 17YF1GA015.

Conflicts of Interest: The authors declare no conflict of interest.

References

1. Yang, Z.; Wu, C.; Zhou, Z.; Zhang, X.; Wang, X.; Liu, Y. Mobility increases localizability: A survey on wireless indoor localization using inertial sensors. *ACM Comput. Surv. (Csur)* **2015**, *47*, 54. [CrossRef]
2. Jyothsna, V.; Prasad, V.V.R.; Prasad, K.M. A review of anomaly based intrusion detection systems. *Int. J. Comput. Appl.* **2011**, *28*, 26–35. [CrossRef]
3. Kausar, F.; Al Eisa, E.; Bakhsh, I. Intelligent home monitoring using RSSI in wireless sensor networks. *Int. J. Comput. Netw. Commun.* **2012**, *4*, 33. [CrossRef]
4. Gansemer, S.; Großmann, U.; Hakobyan, S. Rssi-based euclidean distance algorithm for indoor positioning adapted for the use in dynamically changing wlan environments and multi-level buildings. In Proceedings of the 2010 International Conference on Indoor Positioning and Indoor Navigation (IPIN), Zurich, Switzerland, 15–17 September 2010; pp. 1–6.
5. Dhar, S.; Varshney, U. Challenges and business models for mobile location-based services and advertising. *Commun. ACM* **2011**, *54*, 121–128. [CrossRef]
6. Zhou, Z.; Yang, Z.; Wu, C.; Shangguan, L.; Liu, Y. Omnidirectional coverage for device-free passive human detection. *IEEE Trans. Parallel Distrib. Syst.* **2014**, *25*, 1819–1829. [CrossRef]
7. Youssef, M.; Agrawala, A. The Horus WLAN location determination system. In Proceedings of the 3rd International Conference on Mobile Systems, Applications, and Services, Seattle, WA, USA, 6–8 June 2005; pp. 205–218.
8. Pei, L.; Chen, R.; Liu, J.; Kuusniemi, H.; Tenhunen, T.; Chen, Y. Using inquiry-based Bluetooth RSSI probability distributions for indoor positioning. *J. Glob. Position. Syst.* **2010**, *9*, 122–130.

9. Huang, C.H.; Lee, L.H.; Ho, C.C.; Wu, L.L.; Lai, Z.H. Real-time RFID indoor positioning system based on Kalman-filter drift removal and Heron-bilateration location estimation. *IEEE Trans. Instrum. Meas.* **2015**, *64*, 728–739. [CrossRef]

10. Yasir, M.; Ho, S.W.; Vellambi, B.N. Indoor positioning system using visible light and accelerometer. *J. Lightwave Technol.* **2014**, *32*, 3306–3316. [CrossRef]

11. Zafari, F.; Gkelias, A.; Leung, K. A Survey of Indoor Localization Systems and Technologies. *arXiv*, 2017; arXiv:1709.01015.

12. Zanca, G.; Zorzi, F.; Zanella, A.; Zorzi, M. Experimental comparison of RSSI-based localization algorithms for indoor wireless sensor networks. In Proceedings of the Workshop on Real-World Wireless Sensor Networks, Glasgow, Scotland, 1–4 April 2008; pp. 1–5.

13. Vasisht, D.; Kumar, S.; Katabi, D. Decimeter-Level Localization with a Single WiFi Access Point. In Proceedings of the Networked Systems Design and Implementation, Santa Clara, CA, USA, 16–18 March 2016; pp. 165–178.

14. Huang, X.; Guo, S.; Wu, Y.; Yang, Y. A fine-grained indoor fingerprinting localization based on magnetic field strength and channel state information. *Pervasive Mob. Comput.* **2017**, *41*, 150–165. [CrossRef]

15. Alippi, C.; Vanini, G. A RSSI-based and calibrated centralized localization technique for Wireless Sensor Networks. In Proceedings of the Fourth Annual IEEE International Conference on the Pervasive Computing and Communications Workshops, Pisa, Italy, 13–17 March 2006.

16. Bolliger, P. Redpin-adaptive, zero-configuration indoor localization through user collaboration. In Proceedings of the First ACM International Workshop on Mobile Entity Localization and Tracking in GPS-Less Environments, Istanbul, Turkey, 22–26 August 2008; pp. 55–60.

17. Wu, K.; Xiao, J.; Yi, Y.; Chen, D.; Luo, X.; Ni, L.M. CSI-based indoor localization. *IEEE Trans. Parallel Distrib. Syst.* **2013**, *24*, 1300–1309. [CrossRef]

18. Yang, Z.; Zhou, Z.; Liu, Y. From RSSI to CSI: Indoor localization via channel response. *Acm Comput. Surv. (Csur)* **2013**, *46*, 25. [CrossRef]

19. Li, F.; Al-Qaness, M.A.A.; Zhang, Y.; Zhao, B.; Luan, X. A Robust and Device-Free System for the Recognition and Classification of Elderly Activities. *Sensors* **2016**, *16*, 2043. [CrossRef] [PubMed]

20. Youssef, M.; Mah, M.; Agrawala, A. Challenges: Device-free passive localization for wireless environments. In Proceedings of the ACM International Conference on Mobile Computing and Networking, Montréal, QC, Canada, 9–14 September 2007; pp. 222–229.

21. Wang, J.; Zhang, L.; Wang, X.; Xiong, J.; Chen, X.; Fang, D. A novel CSI pre-processing scheme for device-free localization indoors. In Proceedings of the Eighth Wireless of the Students, by the Students, and for the Students Workshop, New York, NY, USA, 3–7 October 2016; pp. 6–8.

22. Xiao, J.; Wu, K.; Yi, Y.; Ni, L.M. FIFS: Fine-grained indoor fingerprinting system. In Proceedings of the 2012 21st International Conference on Computer Communications and Networks (ICCCN), Munich, Germany, 30 July–2 August 2012; pp. 1–7.

23. Xiao, J.; Wu, K.; Yi, Y.; Wang, L.; Ni, L.M. Pilot: Passive device-free indoor localization using channel state information. In Proceedings of the 2013 IEEE 33rd international conference on Distributed computing systems (ICDCS), Philadelphia, PA, USA, 8–11 July 2013; pp. 236–245.

24. Li, X.; Li, S.; Zhang, D.; Xiong, J.; Wang, Y.; Mei, H. Dynamic-music: Accurate device-free indoor localization. In Proceedings of the 2016 ACM International Joint Conference on Pervasive and Ubiquitous Computing, Heidelberg, Germany, 12–16 September 2016; pp. 196–207.

25. Chapre, Y.; Ignjatovic, A.; Seneviratne, A.; Jha, S. CSI-MIMO: Indoor Wi-Fi fingerprinting system. In Proceedings of the 2014 IEEE 39th Conference on Local Computer Networks (LCN), Edmonton, AB, Canada, 8–11 September 2014; pp. 202–209.

26. Wu, Z.; Xu, Q.; Li, J.; Fu, C.; Xuan, Q.; Xiang, Y. Passive Indoor Localization Based on CSI and Naive Bayes Classification. *IEEE Trans. Syst. Man Cybern. Syst.* **2017**, *48*, 1566–1577. [CrossRef]

27. Wang, X.; Gao, L.; Mao, S. CSI phase fingerprinting for indoor localization with a deep learning approach. *IEEE Internet Things J.* **2016**, *3*, 1113–1123. [CrossRef]

28. Sen, S.; Radunovic, B.; Choudhury, R.R.; Minka, T. You are facing the Mona Lisa: Spot localization using PHY layer information. In Proceedings of the 10th International Conference on Mobile Systems, Applications, and Services, Low Wood Bay, Lake District, UK, 19–25 June 2012; pp. 183–196.

29. Wu, C.; Yang, Z.; Zhou, Z.; Qian, K.; Liu, Y.; Liu, M. PhaseU: Real-time LOS identification with WiFi. In Proceedings of the Computer Communications, Kowloon, Hong Kong, 26 April–1 May 2015; pp. 2038–2046.

30. Ali, A.H.; Razak, M.R.A.; Hidayab, M.; Azman, S.A.; Jasmin, M.Z.M.; Zainol, M.A. Investigation of indoor WIFI radio signal propagation. In Proceedings of the 2010 IEEE Symposium on Industrial Electronics & Applications (ISIEA), Penang, Malaysia, 3–5 October 2010; pp. 117–119.

31. Kemper, J.; Linde, H. Challenges of passive infrared indoor localization. In Proceedings of the Positioning, Navigation and Communication, Hannover, Germany, 27 March 2008; pp. 63–70.

32. Jiang, Y.; Pan, X.; Li, K.; Lv, Q.; Dick, R.P.; Hannigan, M.; Shang, L. Ariel: Automatic wi-fi based room fingerprinting for indoor localization. In Proceedings of the 2012 ACM Conference on Ubiquitous Computing, Pittsburgh, PA, USA, 5–8 September 2012; pp. 441–450.

33. Escudero, G.; Hwang, J.G.; Park, J.G. An Indoor Positioning Method Using IEEE 802.11 Channel State Information. *J. Electr. Eng. Technol.* **2017**, *12*, 1286–1291. [CrossRef]

34. Wu, X.; Kumar, V.; Quinlan, J.R.; Ghosh, J.; Yang, Q.; Motoda, H.; Zhou, Z.H. Top 10 algorithms in data mining. *Knowl. Inf. Syst.* **2008**, *14*, 1–37. [CrossRef]

35. Roos, T.; Myllymäki, P.; Tirri, H.; Misikangas, P.; Sievänen, J. A Probabilistic Approach to WLAN User Location Estimation. *Int. J. Wirel. Inf. Netw.* **2002**, *9*, 155–164. [CrossRef]

36. Qian, K.; Wu, C.; Yang, Z.; Liu, Y.; Zhou, Z. PADS: Passive detection of moving targets with dynamic speed using PHY layer information. In Proceedings of the IEEE International Conference on Parallel and Distributed Systems, Hsinchu, Taiwan, 16–19 December 2015; pp. 1–8.

37. Zhang, L.; Ding, E.; Zhao, Z.; Hu, Y.; Wang, X.; Zhang, K. A novel fingerprinting using channel state information with MIMO–OFDM. *Cluster Comput.* **2017**, *20*, 3299–3312. [CrossRef]

38. Zhuo, Y.; Zhu, H.; Xue, H. Identifying a new non-linear CSI phase measurement error with commodity WiFi devices. In Proceedings of the 2016 IEEE 22nd International Conference on Parallel and Distributed Systems (ICPADS), Wuhan, China, 13–16 December 2016; pp. 72–79.

39. Zheng, L.; Hu, B.; Chen, H. A High Accuracy Time-Reversal Based WiFi Indoor Localization Approach with a Single Antenna. *Sensors* **2018**, *18*, 3437. [CrossRef]

40. Xie, Y.; Li, Z.; Li, M. Precise Power Delay Profiling with Commodity WiFi. In Proceedings of the International Conference on Mobile Computing, Paris, France, 7–11 September 2015; pp. 53–64.

41. Wang, B.; Chen, Q.; Yang, L.T.; Chao, H.C. Indoor smartphone localization via fingerprint crowdsourcing: Challenges and approaches. *IEEE Wirel. Commun.* **2016**, *23*, 82–89. [CrossRef]

42. Wu, C.; Yang, Z.; Liu, Y. Smartphones based crowdsourcing for indoor localization. *IEEE Trans. Mob. Comput.* **2015**, *14*, 444–457. [CrossRef]

43. Hossain, A.K.M.M.; Soh, W.S. A survey of calibration-free indoor positioning systems. *Comput. Commun.* **2015**, *66*, 1–13. [CrossRef]

44. Jung, S.; Lee, S.; Han, D. A crowdsourcing-based global indoor positioning and navigation system. *Pervasive Mob. Comput.* **2016**, *31*, 94–106. [CrossRef]

Article

A Non-Intrusive Approach for Indoor Occupancy Detection in Smart Environments

Bruno Abade, David Perez Abreu * and Marilia Curado

Department of Informatics Engineering, University of Coimbra, Polo II-Pinhal de Marrocos, 3030-290 Coimbra, Portugal; bruno.abade@student.uc.pt (B.A.); marilia@dei.uc.pt (M.C.)
* Correspondence: dabreu@dei.uc.pt

Received: 14 September 2018; Accepted: 13 November 2018; Published: 15 November 2018

Abstract: Smart Environments try to adapt their conditions focusing on the detection, localisation, and identification of people to improve their comfort. It is common to use different sensors, actuators, and analytic techniques in this kind of environments to process data from the surroundings and actuate accordingly. In this research, a solution to improve the user's experience in Smart Environments based on information obtained from indoor areas, following a non-intrusive approach, is proposed. We used Machine Learning techniques to determine occupants and estimate the number of persons in a specific indoor space. The solution proposed was tested in a real scenario using a prototype system, integrated by nodes and sensors, specifically designed and developed to gather the environmental data of interest. The results obtained demonstrate that with the developed system it is possible to obtain, process, and store environmental information. Additionally, the analysis performed over the gathered data using Machine Learning and pattern recognition mechanisms shows that it is possible to determine the occupancy of indoor environments.

Keywords: smart environments; Internet of Things; indoor occupancy; machine learning; data analysis

---

1. Introduction

The Internet of Things (IoT) paradigm enables the interaction between physical objects via application services to add characteristics such as network connectivity, sensing, and actuation allowing to move forward to the Smart Objects approach. Thus, Smart Objects can communicate with each other, share information, and coordinate their actions in order to take smart and cognitive decisions according to the environment where they are deployed [1].

Combining the IoT paradigm and the Smart Objects approach, the concept of Smart City arose. A Smart City uses a variety of sensors and Smart Objects embedded on traditional things and locations (e.g., buildings, parks, and sidewalks) to improve the citizens' quality of life. One of the Smart Cities sectors is Smart Environments, and its definition is given by Cook et al. "A Smart Environment can acquire and apply knowledge about the surroundings and its inhabitants to improve their experience in that ambiance" [2].

Smart Environments have become popular in recent years targeting the automation of everyday tasks in order to improve the quality of life. A typical example of this kind of systems is the management of energy consumption and Heat Ventilation and Air Conditioning (HVAC) [3,4] in Smart Buildings. For the previous particular use case, it is essential to know the occupancy estimation of specific areas in order to trigger the proper actions to minimise consumption during periods of vacancy, optimise ventilation dynamically for occupant comfort, or forecast of long-term behaviors.

To empower the Smart Environment approach, the use of learning mechanisms plays a key role to analyse patterns, predict situations, and take decisions/actions. Thus, new terms such as Ambient

Intelligence (AmI) arise in this context. AmI brings intelligence to our everyday environments, making them sensitive to us. AmI's primary goal is to introduce automation into the environment to generate knowledge about the users and their surroundings, accumulating data and taking smart and cognitive decisions [5]. Making these environments smarter, we can make the life of their occupants simpler and more automated.

A specific research topic framed in the context of a Smart Environment is focused on looking at people, detecting, tracking and identifying them, as a way to offer high-quality, intelligent services, while considering human factors such as life patterns, health, and mood of a person [2]. One example is to analyse patterns of an elderly person and generate an alert when something abnormal happens. For this, knowing the place's occupancy is a priority. Many techniques are developed to detect the presence of people, the most common are cameras and wearable devices. However, these devices suffer from privacy or intrusiveness issues. Research challenges arise with the design of occupancy detection techniques. One of these challenges is related to how to preserve occupants privacy. A Smart Environment system should be designed to avoid identifying occupants or their activities. Thus, there is the need for non-intrusive techniques to detect occupancy or improve the mechanisms already available.

Environmental data are an excellent source of information for occupancy detection since the presence of living beings affects the surroundings through heat or Carbon Dioxide (CO_2) emission without jeopardising the privacy of the occupants in that particular location. Nevertheless, only with data, it is almost impossible to gauge something. Machine Learning (ML) techniques look at the data and try to find patterns; with these patterns, it is possible to affirm the occupancy with a certain percentage of certainty. Although some contributions have been performed in this direction, there is still room for improvement, and this research proposal is focused on that.

The main purpose of this research is the design and development of an affordable and non-intrusive solution to improve occupants experience in Smart Environments with ML support. The proposed solution monitors temperature, light intensity, noise, and CO_2 to estimate the presence of occupants through these environmental features that can be integrated with other existent approaches. First, the data are collected and analysed, before applying ML techniques to infer the occupancy of the area under monitoring. In the first stage, our solution detects the presence or absence of occupants. In the second stage, the number of occupants inside the area of interest is estimated.

This paper is structured as follows. Section 2 discusses the related work. Section 3 presents the solution developed to detect occupants, its architecture, and the key features that were considered, the gathering system and the ML concerns. Section 4 shows the experimental implementation. Section 5 analyses and discusses the results obtained. Finally, conclusions are presented in Section 6 as well as suggestions for future works.

2. Related Work

Occupancy detection systems could be classified according to the need to use a terminal or not [6,7]. In the case of the methods that require a terminal, it is necessary to attach a device to the occupants to keep track of them (e.g., a smartphone). In the non-terminal methods, the detection is based on a passive approach that is focused on monitoring areas or spaces instead of the identification of devices (e.g., cameras monitoring a room). Figure 1 depicts a simple classification of the occupancy detection methods following the terminal and non-terminal approaches, and their more specific characterisations, which are used to organise the discussion of this section.

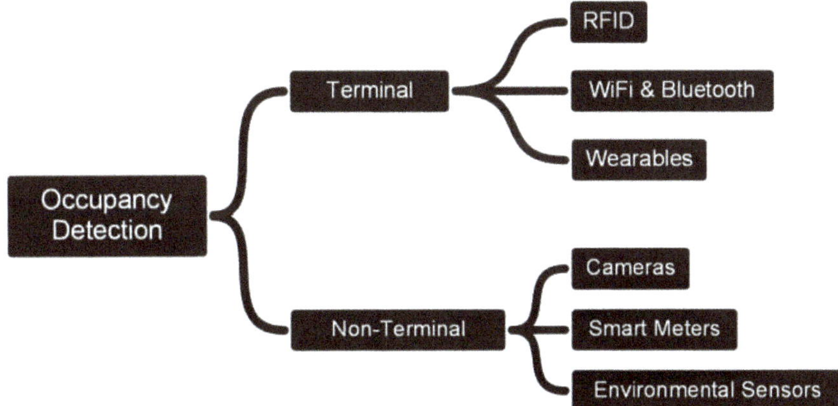

Figure 1. A simple classification of occupancy/location detection methods.

In the branch of terminals, the methods for occupancy and counting lay on devices that have embedded wireless transmitters which support different communication technologies, such as, Radio-Frequency Identification (RFID), WiFi and Bluetooth, or Global Positioning System (GPS) in the case of wearable devices. On the other hand, the branch of non-terminals relies on monitoring specific surroundings by using cameras, smart meters for energy consumption, or environmental sensors (e.g., CO_2 and temperature). A discussion of some relevant works on occupancy detection is presented below.

Hahnel et al. [8] proposed a probabilistic measurement model for RFID readers that allows accurately tracking RFID tags in the environment; specifically, the authors studied the problem of localising the RFID tags using a mobile platform based in robots equipped with RFID antennas. Li and Becerik-Gerber [9] performed a survey of RFID-based solutions and the algorithms used for occupancy and location at indoor environments. After discussing more than twenty projects, authors identify the drawbacks of each solution to move forward to the identification of the most relevant research challenges regarding outdoor/indoor location sensing solutions. In a follow-up research, Li et al. [10] proposed an energy-saving strategy for Smart Buildings based on RFID occupancy detection to support demand-driven HVAC operation by detecting and tracking occupants around areas of interest inside the buildings. The use of RFID technologies for occupancy detection is an affordable option considering the price of receptors and tags; nonetheless, this approach could be affected by electric and magnetic conditions of the environment leading to inaccurate occupancy detection. A more constraining issue is the fact that occupants have to carry a special tag to be monitored, making the process invasive and susceptible to additional errors in case the occupants forget their specific devices somewhere.

Some occupation detection methods take advantage of the communication technologies embedded in devices commonly used by the occupants of the area of interest, such as Smartphones, Smartwatches, and Fitness trackers. Huh and Seo [11] came up with a system that estimates the indoor position of a user taking advantage of some specific characteristics of the Bluetooth protocol. Specifically, the system uses beacon frames to extract information about the Received Signal Strength Indication (RSSI) and trilateration that is processed to infer indoor positioning. Filippoupolitis et al. [12] evaluated how accurate occupancy estimation in indoor environments using Bluetooth Low Energy (BLE) could be in a prototype system composed of BLE beacons, a mobile application, and a server. After performing the analysis of the data collected using three ML approaches (i.e., k-nearest neighbors, Logistic Regression, and Support Vector Machine), the authors concluded that the combination of BLE and ML leads to a good occupancy estimation.

Depatla et al. [13] proposed a framework for counting the total number of people walking in an area based on the WiFi RSSI measurements between a pair of transmitter/receiver antennas. The authors developed a mathematical model to determine the probability distribution of the received signal amplitude as a function of the total number of occupants based on Kullback–Leiber divergence estimation. The results obtained concluded the authors' approach could estimate the total number of people in indoor and outdoor areas with good accuracy. Balaji et al. [14] designed a system, Sentinel, that leverages in the WiFi infrastructure deployed in the area of interest along with Smartphone carried by occupants to estimate occupancy and enhance the performance of the HVAC system via actuation. The Sentinel system proposed by the authors' shows an accuracy of 86%, with 6.2% false negative error regarding the occupancy in indoor environments. Additionally, the tests performed depict that using actuation over the HVAC system it was possible to save around 17.8% energy.

Wearable electronics, such as Smartwatches and Fitness trackers are becoming more ubiquitous and carrying more sensors and communication interfaces. Jin et al. [3] took advantage of the previous statement to investigate the causal influence of user activity on various environmental parameters monitored by occupant-carried multi-purpose sensors. Their results showed that the fusion of the data collected from the sensors available in the wearable devices (e.g., light level, accelerometer, heart rate, Bluetooth, and GPS) achieves a good classification regarding occupancy/location reaching in some cases values around 99% of accuracy. The quality of data obtained using the method that involved wearable devices, WiFi and Bluetooth allows a more accurate occupancy/location estimation; however, these approaches have privacy concerns regarding how to use the data gathered. For example, the use of Bluetooth allows having access to specific and unique information of the devices, such as the MAC address; or in the case of a Fitness tracker the heart rate histogram could reveal some particular condition or disease. This information could be crossed with other data to obtain detailed information about the owner of the device.

In the non-terminal branch for occupancy/location, the methods that use cameras are well-known. Fleuret et al. [15] combined a generative model with dynamic programming to track occlusions and lighting changes in frame images in order to derive the trajectories of each of them. With the proposed model, authors were able to track multiple persons and ranked their trajectories inside the area under study. Alahi et al. [16] addressed the problem of localising people in crowds using a network of heterogeneous cameras by formulating a problem focused on calculating the occupancy vector per each captured frame; this is the discretised occupancy of people on the ground from the foreground silhouettes. The occupancy approach proposed is complemented by a graph-driven tracking procedure suited to deal with the temporal dynamics of people occupancy vectors. The main outcome of this work is a well-defined mathematical formulation to locate people via cameras that record frames with very noisy features. In the same way as with the wearable solutions discussed above, the use of cameras for occupancy/location brings a set of privacy issues regarding the identity of occupants and objects that could represent a problem in the final solution.

A different solution based on Smart Meters is presented by Chen et al. [17]. They tried to predict the occupancy analysing electrical usage. They observed that the home's pattern of electricity usage changes when there are occupants. The study was carried out in two homes and later on correlated with statistical data (e.g., power's mean and variance). Some challenges on non-intrusive occupancy monitoring are also discussed. Another solution by Lee et al. [6] used an array of pyroelectric infrared sensors (PIR) to detect resident's location in a Smart Home. The authors also proposed an algorithm to analyse the information collected from the PIR sensors. The evaluation was carried out using an experimental testbed. Jin et al. [18] tested several binary techniques using data from residential and commercial buildings based on information regarding power usage that requires minimal system calibration and setup, while also ensuring the privacy of the occupants. The accuracy of these works to determine the number of occupants is low and could be considered just an estimation since the power consumption is aggregated, consequently, the exact number of people occupying the area of interest could not be accurate.

The results of Dong et al. [19] indicate that CO_2 and acoustic parameters have the most significant correlation with the number of occupants in a space. Several studies correlate the CO_2 concentrations with the presence of occupants such as in the research of Gruber et al. [20]. Ryu et al. [21] used indoor and outdoor CO_2 concentrations and electricity consumptions of lighting systems in a controlled testbed. Although the CO_2 levels could be used to determine occupancy, gathering this kind of data with good levels of accuracy is not an easy task considering that aspects such as room ventilation, room flow-rate, and presence of plants in the room could drastically influence the concentration and dissipation of the aforementioned gas. Thus, it is not feasible to only use CO_2 as a metric for occupancy estimation.

Candanedo et al. [22] and Amayri [23] estimated occupancy using a combination of heterogeneous sensors. The first research uses data from light, temperature, humidity and CO_2 sensors; and the second one uses data from luminance, temperature, humidity, motion detection, power consumption, and CO_2, as well as data collected from a microphone or door/window burglary sensors. These works follow the same ground truth strategy focused on cameras to corroborate the presence of occupants, which introduce several privacy constraints. Additionally, both works utilise ML techniques to evaluate the results of the proposed solutions, particularly, decision tree learning algorithms. Considering the number of sensors and devices used in these studies were significantly high, our proposal is focused on answering the question whether it is possible to obtain similar or better results regarding occupancy detection using fewer resources. Even more, our proposal uses a non-intrusive ground truth strategy to avoid jeopardising the privacy and security of the occupants in the area of interest.

This research uses as inspiration some of the ideas proposed in the works discussed, in particular, the gathering of data from different sources to move forward to a complete analysis of the data collected using a ML approach. Our goal is to detect occupancy in indoor environments by pre-processing the datasets collected before applying ML to a binary and multi-class problem. Additionally, we design our solution focused on two main requirements: the first one is to try to take advantage of cheap/affordable devices commonly deployed in Smart Environment, and the second one is to guarantee that the privacy of the occupants in the area under study would not be compromised. Thus, after the discussion of the works in this research area, our proposal uses environmental features via the combination of data gathered from different sensors. This solution is presented in the following section.

3. Occupancy Detection in Indoor Environments

As discussed in the previous section, occupancy detection could be used to trigger some actuation mechanisms in Smart Environments in order to improve resource usage and user experience, among other factors. An important issue that must be considered is the preservation of privacy of the data collected and analysed. Additionally, it would be desirable to take advantage of the infrastructure available in the surroundings to avoid incurring in extra expenses, while allowing the scalability of the solution. Considering these factors, this research is focused on a non-intrusive and inexpensive solution for occupancy estimation that ensures occupants privacy while taking advantage of the technological infrastructure already available in common Smart Environments including Smart Buildings and Smart Homes. From the analysis performed on Section 2, and to comply with the previously established requirements, our occupancy detection solution is focused on environmental data.

A scene analysis approach is used in this research to extract the features of interest for indoor scenarios to proceed and then to estimate the occupancy in the area using the gathered data [24]. The scene analysis method does not rely on any theoretical model or specific hardware; however, it requires a preliminary phase for capturing features which are influenced by changes in the area of interest [25].

In this section, we explain the criteria applied to select the features used in our solution before moving forward to the description of the design of the four-layers architecture adopted for the gathering and processing of the data. The section concludes with the discussion of the ML classifiers that were selected to improve the performance of occupancy detection in indoor environments. Table 1 summarises the terms used in the remaining of the manuscript.

Table 1. Notation table.

Term	Meaning
Temp	Temperature
LR	Logistic Regression
SVM	Support Vector Machine
ANN	Artificial Neural Network
TP	True Positive
TN	True Negative
FP	False Positive
FN	False Negative
λ	Regularisation Parameter
pd	Polynomial Degree Parameter
C	Penalty Cost Parameter
γ	Standard Deviation Parameter
hu	Hidden Layers Units

3.1. How Do Human Beings Change Their Surrounding?

A human body is similar to a machine, as to perform actions it needs energy. The first and second laws of thermodynamics state that it is impossible to create energy out of nowhere and a hot body transfers its energy to a cold one. Consequently, the human body is subject to these laws. This energy is interchanged with the environment, in the form of heat, can be by sensible heat (conduction, convection, and radiation) or latent heat transfers (evaporation and condensation) [26]. A healthy adult human releases approximately between 100 Watts (in a resting state) and 1000 Watts (in an effort state), equivalent to the heat dissipated by a few laptops [27]. Thus, the heat of an environment is influenced by the number of persons in there.

Similar to heat, CO_2 is a side effect of the metabolism. It is an essential gas for the existence of life, but at very high concentrations (e.g., greater than 5000 parts per million (ppm)) it can pose a health risk. CO_2 concentrations commonly found in buildings are not a direct health risk, but this concentration can be used as an indicator of occupancy [28]. In fact, occupants are the principal source of CO_2 increasing in indoor environments [29].

In 1879, Thomas Edison invented the first light bulb which made viable to extend the working hours of the human beings [30]. Nowadays, it is common to have artificial light in working spaces. This fact enables the possibility of drawing a relation between light sources and occupancy in indoor environments.

Noise is another feature that is affected by the number of human beings in an ambient [31]. Thus, it is possible to expect that the noise of a specific place will increase with the number of people there. An important fact that should be considered in indoor environments is that they usually have background noise produced by household appliances or other static sources of sound.

Considering the discussion and facts addressed so far in this section, this research uses the following environmental features to detect and estimate the occupancy of indoor environments: heat (via the measurement of the temperature in the area of interest), CO_2, light intensity, and noise. Specifically, a testbed was designed to acquire the data and extract the information to be analysed using a ML approach.

In the next subsection, the architecture designed and used to process the data gathered from the features selected in this work is presented.

3.2. Data Processing Architecture

The data processing architecture used in this research is depicted in Figure 2. The architecture has four layers: Objects Layer, Communication Layer, Analysis Layer, and Application Layer. The functionality of each layer is presented below:

- *Objects Layer*: Deals with the physical sensors that collect raw data information. The sensors used in this research are presented in Table 2.
- *Communication Layer*: Handles the data coming from the sensors. In this layer, the following components are used: an embedded operating system, signal processors, microcontrollers, and gateway nodes. In this layer, the communication between an Arduino Yun (a microcontroller board based on the ATmega32u4 and the Atheros AR9331) and the sensors (e.g., thermistors and sound sensors) is carried out using 10 bits ADC via an I2C bus. The Arduino Yun communicates with a Raspberry Pi (RPi) by Serial Communication performed by the Universal Serial Bus (USB) port to process and store the data gathered.
- *Analysis Layer*: Provides the data management required to extract the necessary information from the raw data collected in the lower layers. This layer includes the elements to perform data mining, analytics services, and device management. The data collected and analysed are stored using a MySQL Database (version 5.5). The tool used to perform the analytics of the data was Matlab (version v.9.2.–R2017a).
- *Application Layer*: Deals with the utilisation of the processed data. It includes services and applications. This last layer uses the previous layers to acquire raw data through sensors, storing and treating it to apply ML techniques to perform the main goal of this research, which is to detect people in indoor environments using a non-intrusive approach.

Table 2. Sensors used in the Objects Layer.

Name	Type	Manufacturing	Communication
NTC Thermistor Module	Temperature	Adafruit	ADC
CCS811 Breakout	CO_2	Adafruit	I2C
Sound Detector	Noise	SparkFun	ADC
TSL2591 Breakout	Light	Adafruit	I2C

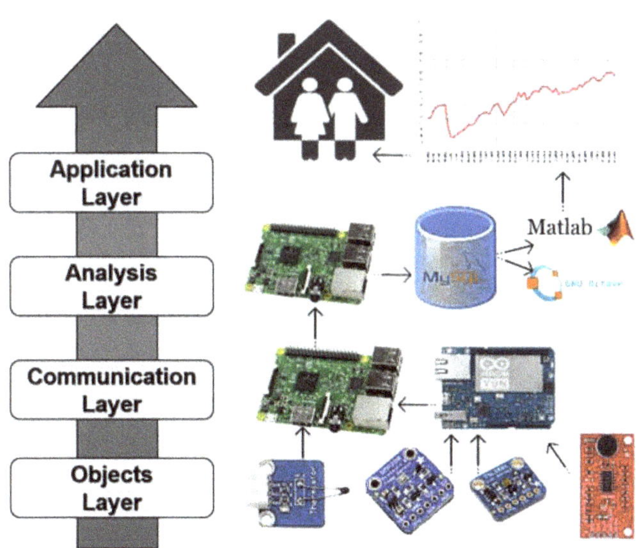

Figure 2. Data processing architecture.

The flow of data in the processing architecture begins at the lower level with the raw data acquisition via environmental sensors. Specifically, every time that the Arduino Yun receives a signal from the RPi (every 10 s), the former one gets ten samples with a delay of 100 microseconds.

Next, an average is computed and sent to the RPi. Finally, a new average is calculated using the aggregated data of six values and it is stored in the MySQL database. With this average, it is possible to decrease the fluctuations in the data. This process is repeated every minute. In the upper layers of the architecture, the analytic functions using ML are run over the data collected and stored after pre-processing it. Thus, the outcome of the Application Layer will be an estimation of the occupancy of the indoor area under study.

3.3. Machine Learning Classifiers and Their Parameters' Tweaking

In the ML context, a supervised approach is used to process the data so the system could learn from it. We use three classifiers: Logistic Regression (LR), a direct probabilistic interpretation; Support Vector Machine (SVM), a hyperplane with the maximal margin to separate the data with similarities; and Artificial Neural Network (ANN), a classifier inspired by how the human brain works. For the ANN case, the hypothesis function is obtained by processing the input features via a set of activation units. These classifiers are largely used in classification problems [32–34]. The purpose of using these three classifiers is to compare and select the best classifier, considering the specific problem and the features involved.

Some parameters can improve the overall performance of the classifier. These parameters have fix/default value results under the same conditions. Thus, it is necessary to tweak the classifiers' parameters according the problem under study to improve the results obtained during the analysis of the data. For LR, the regularisation parameter λ and the degree of the polynomial pd were used. In the case of SVM, we applied the penalty cost C, and the standard deviation parameter γ. For ANN, the number of hidden layers units hu, and the previous λ and pd parameters were utilised.

Regarding the default cases, for LR, a sigmoid function with threshold equal to 0.5, $\lambda = 0$, and $pd = 1$ was used. For SVM, a Radial-Basis Function (RBF) kernel was used in conjunction with these values $C = 1$ and $\gamma = 0$. For ANN, three layers (input layer, hidden layer, and output layer) were used, and the following values were set: $\lambda = 0$, $hu = 1$, and $pd = 1$. These values represent the base cases for each classifier before proceeding with a grid search during the training phase in order to tweak them to improve the performance of each classifier. For LR, the possible values of λ were $[0, 0.01, 0.1, 1, 10, 100]$ and for pd the range was 1–3. For SVM, γ values were $[0, 0.01, 0.1, 1, 10]$, and C could assume $[0.1, 1, 10]$. Finally, for ANN, λ could take the following values $[0, 0.1, 1, 10]$, the range of values of pd was 1–3, and hu could assume 1–3. To select the default and improved values for the classifiers, we used the recommendations of Clarke et al. [32] and Perez et al. [33].

In this research, two problems were studied, a binary problem, where the positive case is the presence of occupants and the negative case is the absence of occupants, and a multi-class problem, where the objective is to determine the exact number of occupants. The research started with the selection of the features to estimate the occupancy in indoor environments, to move forward to the design of the four-layers data processing architecture corresponding to: (1) the sensors to gather the data concerning the features selected; (2) the communication protocols used between sensors, microcontrollers, and processors; (3) how to pre-process and store the data; and (4) the analysis of the data collected using a ML approach where the different classifiers were tested in order to determine the best one by tweaking their parameters.

The setup of the testbed used in this work, as well as how the ML classifiers were evaluated according to their performance is detailed in the next section.

4. Experimental Setting

The experiments were conducted in a room of the Laboratory of Communications and Telematics–Centre for Informatics and Systems located in the middle of the Department of Informatics Engineering at the University of Coimbra. The room has an area of 8.5×5.5 m^2 and 4.15 m of height. This room has a small occupancy change (the maximum number of occupants is five, and the minimum number of occupants is zero) and very low ventilation. The only ventilation in the room is the door and few window cracks. The heating, ventilation, and air conditioning equipment were off during the time of the tests to prevent any influence on the data collected, and it was assumed that the occupants kept the doors and windows closed and the lights on during the period they were in there.

This testbed was set up to study occupancy detection in indoor environments using non-intrusive sensors and ML techniques. The primary objective of the experiments is to evaluate the accuracy regarding occupancy detection in two branches, the simplest one focuses on the presence or absence of occupants; and the more advanced one on the estimation of the number of occupants. In the remainder of this section, we discuss about the placement of the nodes to collect the environmental data and analysis performed to evaluate the accuracy and precision of the classifier utilised. The datasets used in this research, as well as the source code used in the nodes (e.g., Arduinos and RPIs) and the analysis of the data using ML methods, are available via a GitHub repository [35].

4.1. Nodes Placement and Ground Truth Strategy

Three gathering and processing nodes (i.e., 3 RPis and 3 Arduinos) were placed in the room (see Figure 3) in strategic positions to collect data. Figure 4 shows the physical location of the nodes in the area under study. Node 1 has a temperature (in and out) and sound sensors; Node 2 has temperature, CO_2 and sound sensors; and Node 3 has the most significant variety of sensors including temperature, CO_2, sound, and light intensity. Besides gathering environmental data, Node 1 is responsible for controlling the number of occupants in the room (i.e., the ground truth device is attached to it) and also works as the storage node of the data collected during the experiment.

Three CO_2 and temperature sensors were placed on each node, and the average of the values collected were computed to mitigate possible fluctuations. The sound detectors were placed close to the occupants for accuracy purposes. The light sensor was placed as far from the windows as possible so that the main light source incident on it was one of the lamps. Regarding the temperature, a sensor was placed in the hallway and other sensors inside the room. The difference between the temperatures gathered at these two different places was analysed to estimate the occupancy in the room.

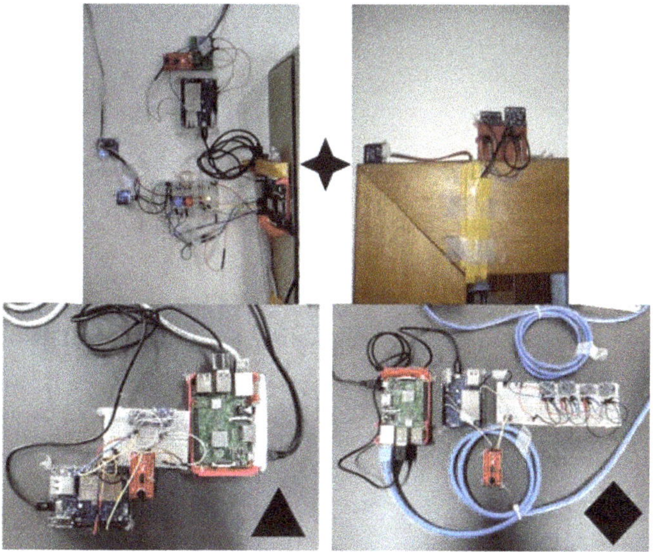

Figure 3. Nodes and sensors deployed in the testbed: (**top**) Node 1 ((**left**) indoor sensors; and (**right**) outdoor temperature sensor); and (**bottom**) Node 2 ((**left**) and Node 3 (**right**).

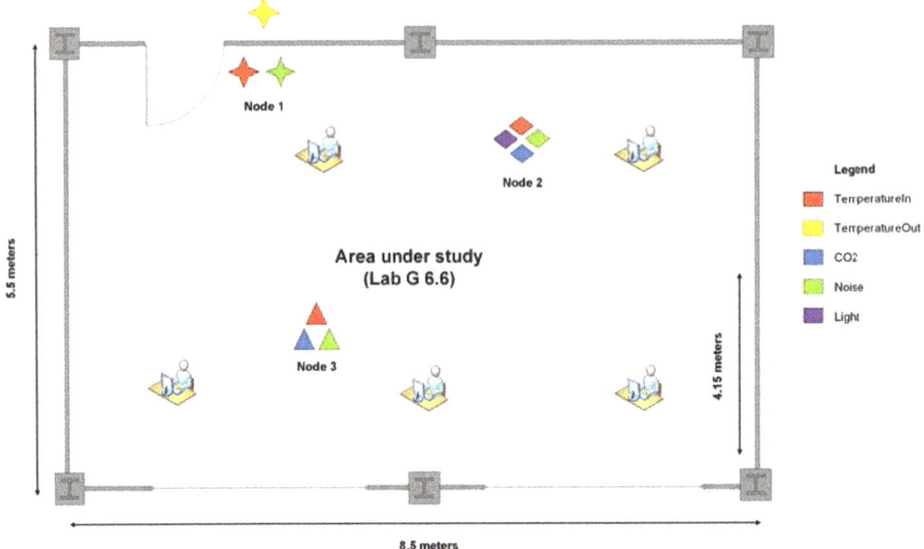

Figure 4. Nodes and sensors placement.

A ground of truth approach was adopted in this research to validate the data gathered. Concretely, in this work, a simple mechanism with two buttons (blue to enter and red to leave) was developed to create the ground truth and register when a person enters or leaves the room. Every time that an occupant presses one of these buttons, the counter is increased or decreased, respectively. To visualise if the number of occupants is correct, three LED were introduced as a binary counter ($2^n - 1$ occupants in the room). The leftmost LED is the most significant and the rightmost LED is the least significant. The number of total occupants by minute is the average of samples acquired every 10 s.

4.2. Classifier Performance Evaluation

To analyse the performance of the classifiers, several criteria were used. Accuracy measures the percentage of entries that were correctly classified (see Equation (1)), and the miss rate measures the percentage of entries that were incorrectly classified (see Equation (2)) [36]. True Positive (TP) and True Negative (TN) represent the correct classification/prediction if the entry belongs to the positive class or negative class, respectively. False Negative (FN) and False Positive (FP) represent the incorrect classification/prediction if the entry does not belong to the negative and positive classes, respectively [36].

$$Accuracy = \frac{TP + TN}{N} * 100 \tag{1}$$

where N is the total size of the training dataset.

$$Missrate = (100 - Accuracy) \tag{2}$$

To evaluate a classifier, it is necessary to verify its accuracy when it has to process new data. The classifiers can have a high accuracy when they are tested with the training dataset, but they can have a low accuracy with a new dataset. Thus, it is recommended to split the data into a training dataset and a testing dataset [37]. The training dataset is suitable to train the classifiers, and the testing dataset is appropriated to measure their performance to new entries. Typically, the dataset is divided into three portions: training (to train the classifiers), cross-validation (to adjust the parameters), and testing (to verify the performance of the classifiers) [33,37]. In this work, the dataset used was split following the same approach, particularly for the cross validation the k-fold method with $k = 5$ was used [38].

In certain cases, the dataset can have skewed classes, i.e., one class has a small set of data. For example, assuming that the training dataset contains 0 positive and 100 negative entries, and if all instances are predicted correctly, the accuracy will be 100%, but the classifier had no chance of learning the hidden patterns. With the previous example, it can be said that the accuracy does not work well when the dataset is unbalanced, i.e., it has more data in one class than in the other.

The F-Score was used to predict the performance of the classifiers. It is a technique that measures the discrimination of classes, through a harmonic mean of two metrics, recall and precision (see Equation (5)) [37]. Recall measures the percentage of entries that belongs to the positive class and was classified/predicted correctly (see Equation (3)) [36]. Precision measures the percentage of hits over the entries of the predicted positive class that really belongs to positive class (see Equation (4)) [36]. To have a high F-Score, both precision and recall must be high.

$$Recall = \frac{TP}{TP + FN} * 100 \tag{3}$$

$$Precision = \frac{TP}{TP + FP} * 100 \tag{4}$$

$$FScore = 2 * \frac{Precision * Recall}{Precision + Recall} \tag{5}$$

Equation (5) can only be applied to binary classification problems, but it can be extrapolated to a multi-class classification problem. The Macro-average method takes the average of precision and recall of each class label (see Equations (6) and (7)) [39,40].

$$Recall = \frac{Recall_1 + Recall_2 + ... + Recall_k}{k} * 100 \tag{6}$$

where k is the class label.

$$Precision = \frac{Precision_1 + Precision_2 + ... + Precision_k}{k} * 100 \tag{7}$$

The analysis and discussion of the results obtained in this research are presented in the next section.

5. Results and Discussion

The data acquisition for this research was performed over two weeks on November 2017 using a rate of one sample per minute. First, the data were analysed and a strategy to use it was defined. It was confirmed that some data had outliers and noise; consequently, to mitigate this issue, two filters (i.e., an outlier filter and a Low-Pass Filter (LPF)) were applied. The performance of the filters over the data is depicted in Figure 5.

An LPF is a circuit that offers easy passage to low-frequency signals and difficult passage to high-frequency signals. Equation (8) gives the discrete implementation of the first order LPF, where α is the smoothing factor, y is the filtered output, x is the input, and n is the sample index. Calculating the next value through this smoothing factor and the previous value, it was possible to reduce the data noise, making the transitions between samples slower and smoother.

$$y[n] = \alpha x[n] + (1 - \alpha)y[n - 1] \tag{8}$$

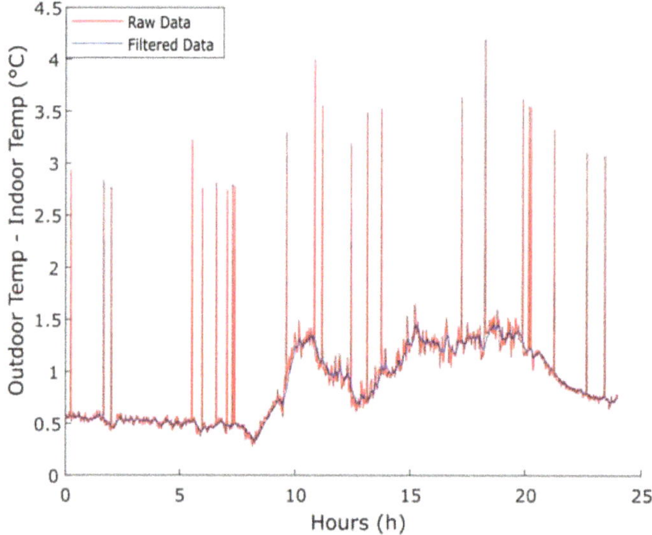

Figure 5. Performance of the outlier and LPF filters over the temperature (i.e., difference of outdoor and indoor temperatures) data gathered.

In a second stage, ten consecutive days of data were selected, representing a total of 14,400 samples (where almost 25% represented positive cases, and 75% represented negative cases) for each dataset and the ML mechanisms were applied. The dataset was divided into two portions, training and testing. The training portion was then subdivided into two portions, training and cross-validation portions, respectively, representing 70% of the original dataset and the remaining 30% corresponds to a testing portion. Within the first portion (i.e., training) 80% was used for training and 20% was used for cross-validation according to the k-fold (with a $k = 5$) approach to prevent overfitting [38]. This value for $k = 5$ was selected given the unbalanced nature of our dataset, where the average occupancy was around 8 h per day that corresponds to office hours; thus, having periods of 16 h without relevant data per day. Using this value (i.e., $k = 5$), we minimise the probability of having k-portions without any relevant data.

In the remaining of the section, a discussion about how the data were pre-processed and the approach used during the binary and multi-class problems is presented.

5.1. Data Pre-Processing

The processing and analysis of the environmental data gathered during the research are depicted in Figure 6a–d. In the charts, the blue and red lines represent a day without and with occupants in the area under study, respectively. Particularly, in Figure 6b, the red line depicts a day with precisely one occupant in the room and the yellow line a day with more than one occupant.

In Figure 6a, the graph is in Celsius degrees by hours. For this analysis, it is important to point out that the data corresponds to the subtraction of the indoor and outdoor temperature as it was described in Section 4.1. For the data collected, it is possible to conclude that the temperature's difference is higher with occupants than without occupants. The first occupant arrived around 09:00 and the last occupant left around 18:00. There are a couple of exceptions around 10:00 and around 12:00. The first one happens because of the incidence of the sunlight in the room, which on this period of test occurs at this hour, increasing the indoor temperature. The second one occurs when the occupants left the room to have lunch.

In Figure 6b, the graph is in ppm per hour. It is possible to see that when an occupant arrived (i.e., around 09:00) the CO_2 levels increased approximately 500 ppm. This increase was more noticeable when more than one occupant was in the room, increasing to around 2000 ppm. In days without occupants, the levels were between 400 and 450 ppm.

Figure 6c depicts the noise data processed. It is possible to see that the differences are not significant considering that the values are almost the same with or without people.

The light intensity data is depicted in Figure 6d. It is possible to see that when an occupant arrived, close to 10:00, the lux increased to around 110 and when he left, close to 18:00, the lux decreased to zero. Around 10:00, it is possible to notice an increase in the light intensity as a result of the incidence of the sunlight in the room. This behavior is consistent with the results and the observations performed during the analysis of the temperature in the room.

After pre-processing the data, each classifier was tested. The results obtained are presented in the next two subsections.

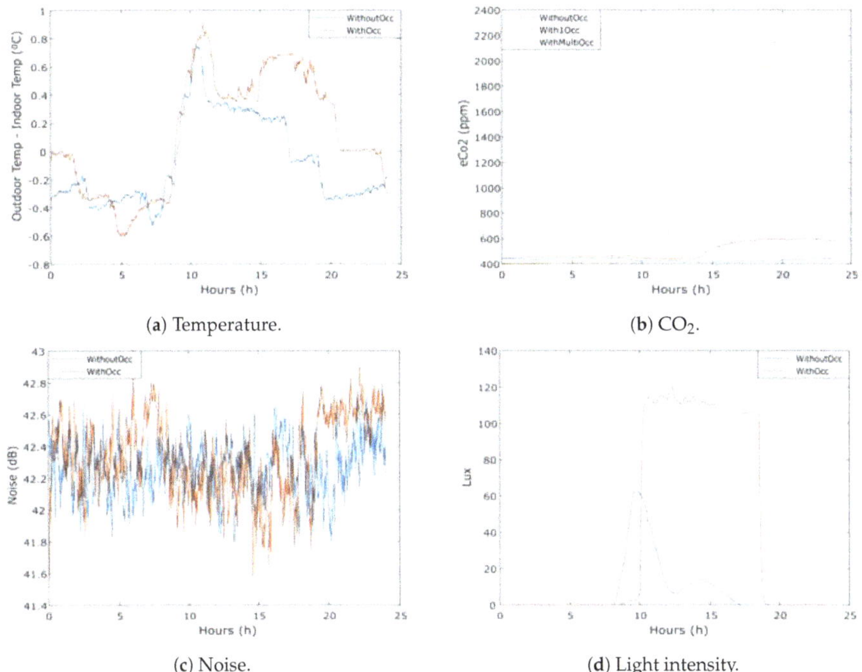

(a) Temperature.

(b) CO_2.

(c) Noise.

(d) Light intensity.

Figure 6. Processing and analysis of the environmental data gathered.

5.2. Binary Problem

The binary problem aims to determine whether an occupant was in the room ($y = 1$) or not ($y = 0$). Table 3 presents the results by applying the classifiers with the dataset without changing the parameters. Analysing the results, the classifiers with best F-Scores were LR followed by SVM. The ANN classifier had the lowest F-Score in almost all the cases. In some cases, the result was 0%, i.e., the classifier could not predict any positive outcome.

Table 3. F-Score results of applying ML algorithms to the data collected for the binary problem with default parameters.

	LR	SVM	ANN
Temp	89.70%	89.66%	89.60%
CO_2	6.59%	1.43%	0%
Noise	1%	1.28%	0%
Light	95.60%	95.60%	95.42%

Regarding the noise as an element to estimate occupancy, the F-Score results show that for the area under study this feature is not a good indicator; considering that the room is a workplace, people are usually concentrated and spend most of the time in silence. The results of the F-Score related to the light intensity were satisfactory with the limitation that this approach could not be used to estimate the number of occupants in the room, just their presence.

As more occupants in the room usually results in higher CO_2 concentrations, these data can detect the number of occupants, as can be seen in Figure 6b. However, because the room does not have a good air flow rate, this concentration reduces slowly and can take hours to stabilise. Consequently, this approach to estimate the number of occupants inside the laboratory did not perform as expected. One possible approach to enhance the results obtained would be to calculate the derivative and then check whether it has a certain slope to determine if an occupant arrived or left the room; nevertheless, this method requires more data, as well as more analysis.

The temperature data suffer from the same problem than the CO_2 data. It is difficult to have a fixed number of occupants in the room. Thus, it was important to have a dataset with more data for calculating the time taken for the temperature to stabilise to improve the results. However, even without this knowledge, the results were satisfactory (i.e., we obtained in average an F-Score of 89%) to detect the presence of occupants.

Table 4 presents the parameters for which the highest F-Scores were obtained using the LR, SVM, and ANN classifiers to the dataset collected. When performing a new F-Score and changing the parameters and the polynomial degree, some features show an improvement, such as CO_2 with an enhancing on the F-Scores of the classifiers between 15% and 47%. Light, temperature, and noise did not show significant growth. Even though the CO_2 levels can be used to infer the number of occupants, the data analysed has to suffer changes before applying a ML technique. The noise had a low F-Score, and the light indicated only the presence or absence of occupants. For these reasons, only the temperature was analysed in a multi-class problem.

Table 4. F-Score results of parameters that perform the highest score for LR, SVM, and ANN for the binary problem.

	LR			SVM			ANN			
	λ	*pd*	F-Score	γ	C	F-Score	λ	*pd*	*hu*	F-Score
Temp	0	2	89.80% (+0.10%)	0.1	1	89.71% (+0.05%)	0.1	1	2	89.72% (+0.12%)
CO_2	0	3	22.10% (+15.51%)	10	10	43.98% (+42.55%)	0	1	2	47.81% (+47.81%)
Noise	0	2	2.60% (+1.60%)	10	10	4.17% (+2.89%)	0	1	3	0% (+0.00%)
Light	10	1	95.60% (+0.00%)	1	0.1	95.55% (+0.13%)	1	1	1	95.32% (−0.10%)

The F-Scores reached by the classifiers, particularly in the case of Temp and Light, show that it is possible to obtain high accuracy regarding indoor occupancy using the LR, SVM, and ANN classifiers, which is aligned with some of the results reported in the state of the art. Specifically, in the research work of Candanedo et al. [22], a research framed within the same topic although using a different dataset, the authors obtained an accuracy of 85.33% for Temp and 97.66% for Light using a Linear Discriminant Analysis (LDA) model. These results are comparable with the values obtained by LR and SVM in this research considering the classifiers' linearity. In the case of Temp, LR and SVM showed better accuracy (i.e., around 4%). On the other hand, for Light, LDA performed better than LR and SVM by around 2%. Instead of an ANN approach, Candanedo et al. [22] decided to determine the performance of Classification and Regression Trees (CART) learning algorithms for this specific problem. The accuracy results obtained using CART were 86.51% and 99.31% for Temp and Light, respectively. Thus, for Temp, the ANN approach had a better accuracy 89.72% (i.e., around 3%), and, in the case of Light, the CART model beat the ANN classifier by around 4% (i.e., 99.31% against 95.32%).

5.3. Multi-Class Problem

The multi-class problem aims to estimate the number of occupants in a room. During this work, on average, there were five occupants in the room. After observing the behavior of the data gathered and the binary problem results, it is possible to conclude that temperature is the more interesting feature to be tested in the multi-class approach. Even though it was possible to obtain F-Scores beyond 95% for all the classifiers for the data corresponding to light, its binary nature makes impossible to use it to estimate the number of occupants in the area under study, thus it was discarded.

Table 5 summarises both the parameters and the F-Scores obtained using the default values for said parameters, and after tweaking them. Concerning the analysis of the temperature data using the default parameters for each classifier, the following F-Scores results were obtained: 24.43% for LR, 24.90% for SVM, and 25.15% for ANN. These results are far away from what we had expected, thus an additional tweaking of the parameter was applied in order to improve the F-Scores. In the best scenario, the following F-Scores results were obtained: 29.43% (more 5%) for LR; 29.72% (more 4.82%) for SVM; and 28.70% (more 3.55%) in the case of ANN.

The results obtained for the multi-class problem show that it was not possible to estimate the number of occupants using just the temperature data. When the default parameters were used, all the classifiers reported almost the same F-Scores, i.e., around 24.5%. After changing the parameters, the SVM classifier produced the best results for this dataset, around 29.72%. It was also assumed that the human body surface had a uniform temperature and a consistent heat production, but this is not necessarily true. The human body has a distinct physical shape and also has multiple thermo-physiological properties. Thus, it is difficult to include those factors in a numerical constant in an indoor space.

Table 5. F-Score results of parameters for LR, SVM, and ANN for the multi-class problem.

	LR			SVM			ANN			
	λ	*pd*	F-Score	γ	C	F-Score	λ	*pd*	*hu*	F-Score
Default	0	1	24.43%	0	1	24.90%	0	1	1	25.15%
Tweaked	0.01	2	29.43% (+5%)	10	10	29.72% (+4.82%)	0	1	3	28.70% (+3.55%)

A valid estimation of the number of occupants in indoor environments using non-intrusive environmental sensors requires a deep study of the correlation of the data gathered. ML techniques have proven to be useful to better understand the interaction and behavior between the sensors, according to the changes induced by the occupants in indoor environments. For the multi-class problem, the correlation between light, temperature, and CO_2 looks promising. In a first step, the analysis of the light could determine accurately the presence or absence of occupants, and, as a second step, a study of the correlation between temperature and CO_2 could enhance the estimation concerning the number of occupants in a specific area. Thus, these open issues lead to the possibility to perform future research in this field.

6. Conclusions

Nowadays, companies and researchers are working on enhancing the quality of life of citizens, using the IoT paradigm to reach the idea of building Smart Environments. In this context, it would be beneficial to have mechanisms to predict or estimate the occupancy of indoor environments to make smart decisions about how to self-adapt to the environmental conditions.

In this research, a solution for occupancy detection with non-intrusive devices using sensors such as temperature, noise, CO_2, and light intensity was proposed and tested. A functional system, made up of a device to gather and process environmental data, and to analyse the data patterns over the collected data regarding people occupancy in indoor environments using ML methods, was tested. The analysis performed allows asserting that with features such as noise data in working environments the performance of the recognition system might be degraded. However, with features such as temperature, CO_2, and light data, it will be possible to estimate the detection of occupants with an acceptable level of accuracy. Thus, the work done in this research could feed third-party applications focused on indoor occupancy detection to generate smart decisions considering the occupants' needs.

For future works, it is necessary to study the full correlation of the environmental data used in this research. A starting point could be the analysis of features and their impact on the system. The CO_2 data have to endure some processing to find the most meaningful way to use this type of data, since it was found that they represent one of the best features to detect the number of occupants.

Author Contributions: B.A. designed the methodology. He also analysed the data, interpreted results and wrote the paper. D.P.A. participated in the research design and writing the paper. M.C. conceived the study and reviewed the manuscript critically for important intellectual content. All authors read and approved the final manuscript.

Funding: This work was financed by national funding via the Foundation for Science and Technology and by the European Regional Development Fund (FEDER), through the COMPETE 2020–Operational Program for Competitiveness and Internationalization (POCI). Additionally, the work presented in this paper was partially carried out in the scope of the MobiWise project: From mobile sensing to mobility advising (P2020 SAICTPAC/0011/2015), co-financed by COMPETE 2020, Portugal 2020–Operational Program for Competitiveness and Internationalization (POCI), European Union s ERDF (European Regional Development Fund), and the Portuguese Foundation for Science and Technology (FCT).

Acknowledgments: David Perez Abreu wishes to acknowledge the Portuguese funding institution FCT-Foundation for Science and Technology for supporting his research under the Ph.D. grant ≪SFRH/BD/117538/2016≫.

Conflicts of Interest: The authors declare that they have no competing interests.

References

1. Atzori, L.; Iera, A.; Morabito, G. The Internet of Things: A survey. *Comput. Netw.* **2010**, *54*, 2787–2805. [CrossRef]

2. Cook, D.; Das, S. How smart are our environments? An updated look at the state of the art. *Pervasive Mob. Comput.* **2007**, *3*, 53–73. [CrossRef]

3. Jin, M.; Zou, H.; Weekly, K.; Jia, R.; Bayen, A.M.; Spanos, C.J. Environmental sensing by wearable device for indoor activity and location estimation. In Proceedings of the IECON 2014—40th Annual Conference of the IEEE Industrial Electronics Society, Dallas, TX, USA, 29 October–1 November 2014; IEEE: Dallas, TX, USA, 2014; Volume 1, pp. 5369–5375. [CrossRef]

4. Imanishi, T.; Tennekoon, R.; Palensky, P.; Nishi, H. Enhanced building thermal model by using CO_2 based occupancy data. In Proceedings of the IECON 2015—41st Annual Conference of the IEEE Industrial Electronics Society, Yokohama, Japan, 9–12 November 2015; Volume 1, pp. 3116–3121. [CrossRef]

5. Cook, D.; Augusto, J.; Jakkula, V. Ambient intelligence: Technologies, applications, and opportunities. *Pervasive Mob. Comput.* **2009**, *5*, 277–298. [CrossRef]

6. Lee, S.; Ha, K.N.; Lee, K.C. A pyroelectric infrared sensor-based indoor location-aware system for the smart home. *IEEE Trans. Consum. Electron.* **2006**, *52*, 1311–1317. [CrossRef]

7. Labeodan, T.; Zeiler, W.; Boxem, G.; Zhao, Y. Occupancy measurement in commercial office buildings for demand-driven control applications—A survey and detection system evaluation. *Energy Build.* **2015**, *93*, 303–314. [CrossRef]

8. Hahnel, D.; Burgard, W.; Fox, D.; Fishkin, K.; Philipose, M. Mapping and localization with RFID technology. In Proceedings of the ICRA '04 2004 IEEE International Conference on Robotics and Automation, New Orleans, LA, USA, 26 April–1 May 2004; Volume 1, pp. 1015–1020. [CrossRef]

9. Li, N.; Becerik-Gerber, B. Performance-based evaluation of RFID-based indoor location sensing solutions for the built environment. *Adv. Eng. Inf.* **2011**, *25*, 535–546. [CrossRef]

10. Li, N.; Calis, G.; Becerik-Gerber, B. Measuring and monitoring occupancy with an RFID based system for demand-driven HVAC operations. *Autom. Constr.* **2012**, *24*, 89–99. [CrossRef]

11. Huh, J.H.; Seo, K. An Indoor Location-Based Control System Using Bluetooth Beacons for IoT Systems. *Sensors* **2017**, *17*, 2917. [CrossRef] [PubMed]

12. Filippoupolitis, A.; Oliff, W.; Loukas, G. Bluetooth Low Energy Based Occupancy Detection for Emergency Management. In Proceedings of the 2016 15th International Conference on Ubiquitous Computing and Communications and 2016 International Symposium on Cyberspace and Security (IUCC-CSS), Granada, Spain, 14–16 December 2016; Volume 1, pp. 31–38. [CrossRef]

13. Depatla, S.; Muralidharan, A.; Mostofi, Y. Occupancy Estimation Using Only WiFi Power Measurements. *IEEE J. Sel. Areas Commun.* **2015**, *33*, 1381–1393. [CrossRef]

14. Balaji, B.; Xu, J.; Nwokafor, A.; Gupta, R.; Agarwal, Y. Sentinel: Occupancy Based HVAC Actuation Using Existing WiFi Infrastructure Within Commercial Buildings. In Proceedings of the SenSys '13 11th ACM Conference on Embedded Networked Sensor Systems, Roma, Italy, 11–15 November 2013; ACM: New York, NY, USA, 2013. [CrossRef]

15. Fleuret, F.; Berclaz, J.; Lengagne, R.; Fua, P. Multicamera People Tracking with a Probabilistic Occupancy Map. *IEEE Trans. Pattern Anal. Mach. Intell.* **2008**, *30*, 267–282. [CrossRef] [PubMed]

16. Alahi, A.; Jacques, L.; Boursier, Y.; Vandergheynst, P. Sparsity Driven People Localization with a Heterogeneous Network of Cameras. *J. Math. Imaging Vis.* **2011**, *41*, 39–58. [CrossRef]

17. Chen, D.; Barker, S.; Subbaswamy, A.; Irwin, D.; Shenoy, P. Non-Intrusive Occupancy Monitoring Using Smart Meters. In Proceedings of the 5th BuildSys'13 ACM Workshop on Embedded Systems for Energy-Efficient Buildings, Roma, Italy, 11–15 November 2013; ACM: New York, NY, USA, 2013; pp. 1–8. [CrossRef]

18. Jin, M.; Jia, R.; Spanos, C.J. Virtual Occupancy Sensing: Using Smart Meters to Indicate Your Presence. *IEEE Trans. Mob. Comput.* **2017**, *16*, 3264–3277. [CrossRef]

19. Dong, B.; Andrews, B.; Lam, K.P.; Höynck, M.; Zhang, R.; Chiou, Y.S.; Benitez, D. An information technology enabled sustainability test-bed (ITEST) for occupancy detection through an environmental sensing network. *Energy Build.* **2010**, *42*, 1038–1046. [CrossRef]

20. Gruber, M.; Trüschel, A.; Dalenbäck, J.O. CO_2 sensors for occupancy estimations: Potential in building automation applications. *Energy Build.* **2014**, *84*, 548–556. [CrossRef]

21. Ryu, S.H.; Moon, H.J. Development of an occupancy prediction model using indoor environmental data based on machine learning techniques. *Build. Environ.* **2016**, *107*, 1–9. [CrossRef]

22. Candanedo, L.; Feldheim, V. Accurate occupancy detection of an office room from light, temperature, humidity and CO_2 measurements using statistical learning models. *Energy Build.* **2016**, *112*, 28–39. [CrossRef]

23. Amayri, M.; Arora, A.; Ploix, S.; Bandhyopadyay, S.; Ngo, Q.D.; Badarla, V.R. Estimating occupancy in heterogeneous sensor environment. *Energy Build.* **2016**, *129*, 46–58. [CrossRef]

24. Liu, H.; Darabi, H.; Banerjee, P.; Liu, J. Survey of Wireless Indoor Positioning Techniques and Systems. *IEEE Trans. Syst. Man Cybern. Part C (Appl. Rev.)* **2007**, *37*, 1067–1080. [CrossRef]

25. Papapostolou, A.; Chaouchi, H. Scene analysis indoor positioning enhancements. *Ann. Telecommun.* **2011**, *66*, 519–533. [CrossRef]

26. Parsons, K. *Human Thermal Environments: The Effects of Hot, Moderate, and Cold Environments on Human Health, Comfort, and Performance*, 3rd ed.; CRC Press: Boca Raton, FL, USA, 2014.

27. Paradiso, J. Systems for Human-powered Mobile Computing. In Proceedings of the DAC '06, 43rd Annual Design Automation Conference, San Francisco, CA, USA, 24–28 July 2006; ACM: New York, NY, 2006; pp. 645–650. [CrossRef]

28. AINSI/ASHRAE. *AINSI/ASHRAE Standard 62.1—2013 Ventilation for Acceptable Indoor Air Quality*; Standard; ASHRAE: Washington, DC, USA, 2013.

29. You, Y.; Niu, C.; Zhou, J.; Liu, Y.; Bai, Z.; Zhang, J.; He, F.; Zhang, N. Measurement of air exchange rates in different indoor environments using continuous CO_2 sensors. *J. Environ. Sci.* **2012**, *24*, 657–664. [CrossRef]

30. Chepesiuk, R. Missing the dark: Health effects of light pollution. *Environ. Health Perspect.* **2009**, *117*, 20–27. [CrossRef] [PubMed]

31. Jensen, K.; Arens, E.; Zagreus, L. Acoustical Quality in Office Workstations, as Assessed by Occupant Surveys. In Proceedings of the 10th International Conference on Indoor Air Quality and Climate (Indoor Air 2005), Beijing, China, 4–9 September 2005; Volume 1, pp. 2401–2405.

32. Clarke, B.; Fokoue, E.; Zhang, H.H. *Principles and Theory for Data Mining and Machine Learning*, 1st ed.; Springer Science & Business Media: Berlin, Germany, 2009.

33. Perez, D.; Astor, M.; Abreu, D.P.; Scalise, E. Intrusion detection in computer networks using hybrid machine learning techniques. In Proceedings of the 2017 XLIII Latin American Computer Conference (CLEI), Cordoba, Argentina, 4–8 September 2017; Volume 1, pp. 1–10. [CrossRef]

34. Domingos, P. A Few Useful Things to Know About Machine Learning. *Commun. ACM* **2012**, *55*, 78–87. [CrossRef]

35. Abade, B. Abade002/A-non-intrusive-approach-for-indoor-occupancy-detection-in-Smart-Environments: Second Release! *Abade002* **2018**. [CrossRef]

36. Powers, D. Evaluation: From precision, recall and fmeasure to roc, informedness, markedness and correlation. *J. Mach. Learn. Technol.* **2011**, *2*, 37–63.

37. Bishop, C. *Pattern Recognition and Machine Learning (Information Science and Statistics)*, 1st ed.; Springer: New York, NY, USA, 2007.

38. Rodriguez, J.D.; Perez, A.; Lozano, J.A. Sensitivity Analysis of k-Fold Cross Validation in Prediction Error Estimation. *IEEE Trans. Pattern Anal. Mach. Intell.* **2010**, *32*, 569–575. [CrossRef] [PubMed]

39. Baldi, P.; Brunak, S.; Chauvin, Y.; Andersen, C.A.; Nielsen, H. Assessing the accuracy of prediction algorithms for classification: an overview. *Bioinformatics* **2000**, *16*, 412–424. [CrossRef] [PubMed]

40. Sebastiani, F. Machine Learning in Automated Text Categorization. *ACM Comput. Surv.* **2002**, *34*, 1–47. [CrossRef]

Article

Computational Efficiency-Based Adaptive Tracking Control for Robotic Manipulators with Unknown Input Bouc–Wen Hysteresis

Kan Xie [1,2], **Yue Lai** [1,3] **and Weijun Li** [1,4]*

[1] School of Automation, Guangdong University of Technology, Guangzhou 510006, China; kanxiegdut@gmail.com (K.X.); Hot_day@163.com (Y.L.)

[2] Guangdong Key Laboratory of IoT Information Technology, Guangzhou 510006, China

[3] Key Laboratory of Ministry of Education, Guangzhou 510006, China

[4] State Key Laboratory of Precision Electronic Manufacturing Technology and Equipment, Guangzhou 510006, China

* Correspondence: weijunli@gdut.edu.cn; Tel.: +86-20-3932-2552

Received: 7 May 2019; Accepted: 15 June 2019; Published: 20 June 2019

Abstract: In order to maintain robotic manipulators at a high level of performance, their controllers should be able to address nonlinearities in the closed-loop system, such as input nonlinearities. Meanwhile, computational efficiency is also required for real-time implementation. In this paper, an unknown input Bouc–Wen hysteresis control problem is investigated for robotic manipulators using adaptive control and a dynamical gain-based approach. The dynamics of hysteresis are modeled as an additional control unit in the closed-loop system and are integrated with the robotic manipulators. Two adaptive parameters are developed for improving the computational efficiency of the proposed control scheme, based on which the outputs of robotic manipulators are driven to track desired trajectories. Lyapunov theory is adopted to prove the effectiveness of the proposed method. Moreover, the tracking error is improved from ultimately bounded to asymptotic tracking compared to most of the existing results. This is of important significance to improve the control quality of robotic manipulators with unknown input Bouc–Wen hysteresis. Numerical examples including fixed-point and trajectory controls are provided to show the validity of our method.

Keywords: sensing and control; computational efficiency; robotic manipulators; hysteresis; adaptive control

---

1. Introduction

It is well-known that robotic manipulators are a class of important systems in industrial and academic research [1]. Based on their widespread use in engineering fields, the control of robotic manipulators has attracted much attention of researchers of robotic systems and control science [2–8]. The modern demand for electronics requires robotic manipulators to be operated in a high-demanding status to reject possible nonlinearities in the closed-loop systems. One of the current research topics is to investigate unknown input nonlinearities in the robotic manipulators.

In practical systems, control inputs are one of the essential units in the closed-loop system and play a key role in maintaining performance and quality [9]. As for the nonlinearities on the input signal, backlash nonlinearity is considered for output feedback control of uncertain nonlinear systems in [10] through backlash inverse. Fu and Xie [11] considered a quantized control problem using a sector bound approach and quantized output feedback systems using a dynamic scaling method [12]. A system with a hysteretic quantizer is considered by Hayakawa et al. [13] to cancel the chattering caused by the logarithmic quantizer. Zhou et al. [14] considered a quantization control problem in a class of systems with parameterized uncertainties and handled it using an adaptive backstepping-based design approach. External disturbances and unknown input nonlinearities are considered for multi-agent systems in [15] and for distributed control of heterogeneous multi-agent systems [16]. Xie et al. [17] addressed unknown input quantization for nonlinear systems and proposed an asymptotic neural-network-based control method. Most recently, faults on control inputs and sensors in multi-agent systems are considered in [18]. Cao et al. [19] used Madelung's rules to propose a method to model symmetric hysteresis. Furthermore, this method is translated into an algorithm that can be run by digital processors. Hysteresis nonlinearity was considered in [20] for decentralized stabilization of interconnected systems. Later, hysteresis inverse was given in [21] for adaptive output feedback control. It is noted that the parameters of hysteresis in [21] must be available for the control design. The work of [22] considered the tracking control of a magnetic shape memory actuator by combining the modeling technique of an inverse Preisach model and sliding mode control design. The work of [23] studied both time delay and actuator saturation in the formation control of teleoperating systems, which cover robotic systems. The works of [24,25] considered hysteresis nonlinearities in the systems and proposed computational-efficiency-based modeling methods to efficiently and precisely describe hysteresis characteristics.

One has to seek extra controls to handle unknown input coefficients and extra disturbances brought from unknown hysteresis for applications in electronics-based systems. To handle the unknown input coefficients, a Nussbaum function-based control method [26] is considered in the literature [27–31]. The work of [32] used the Nussbaum function for a class of single-input single-output systems. Based on [32], the work of [33] considered unknown control coefficients and model uncertainties and provided a robust control for the segway. Note that single-input single-output systems are not feasible for most of the robotic manipulators, where joint spaces should have six dimensions. Thus, for multiple-input multiple-output systems, some works [34–37] are provided to handle a group of multiple Nussbaum functions by different control strategies. Most recently, the work of [38] proposed a Nussbaum function with saturated property and used it to address unknown input nonlinearities in robotic systems, with a focus on eliminating the control shock from Nussbaum functions. The work of [39] considered the intrusion detection problem in underwater wireless networks.

Motivated by the above analysis and the technique on the elimination of overparametrization [40], we combine the adaptive control technique and a dynamical gain-based approach to address unknown input Bouc–Wen hysteresis for robotic manipulators. We model the input hysteresis and integrate it with robotic manipulators. Then, two adaptive mechanisms are proposed in our control scheme. Note that computational efficiency is one of the important issues for the implementation of robotic manipulators. We consider such an issue by proposing a control scheme based on two adaptive laws. One adaptive law is used to handle unknown parameter vectors associated with the regression matrix. The other one is used to address input hysteresis and to allow parameters of the hysteresis in each channel of inputs to be different. The two-parameter control scheme plays a key role in improving computational efficiency for potential real-time implementation. A Lyapunov-method-based stability is given to prove the effectiveness of the proposed adaptive scheme. It is shown that even in the presence of unknown input Bouc–Wen hysteresis, the trajectory tracking objective is ensured for robotic manipulators. Moreover, the tracking error is set to be asymptotic within our adaptive control, while most existing results are ultimately bounded. The asymptotic control derived from our method is of importance to high-demanding applications such as manufacturing.

The remaining parts of this paper are organized as follows. We define an unknown input Bouc–Wen hysteresis control problem for robotic manipulators in Section 2. In Section 3, we present a solution containing two adaptive parameters to address the control problem using a dynamical gain-based approach. In Section 4, simulation studies, including fixed-point control and trajectory control, are presented to validate the method's effectiveness. We summarize the obtained results in Section 5.

2. Problem Formulation

A class of robotic manipulators, such as robotic manipulators, are formulated as the following differential equation [1–4]:

$$D(q)\ddot{q} + H(q,\dot{q})\dot{q} + W(q) = v, \tag{1}$$

where $q \in \mathbb{R}^{L\times 1}$ is a system state vector, $D(q) \in \mathbb{R}^{L\times L}$ denotes an inertial matrix, $H(q,\dot{q}) \in \mathbb{R}^{L\times L}$ represents the Coriolis and centrifugal matrix of the ith robotic arm, $W(q) \in \mathbb{R}^{L\times 1}$ denotes the gravitational force vector, $v \in \mathbb{R}^{L\times 1}$ means the input of the robotic manipulator and will drive the joint space variable q to a predetermined trajectory. Note that the robotic manipulator governed by (1) is capable of modeling jet engines and aircraft.

Here, we specify the input hysteresis nonlinearities as

$$v = \omega(u(t),t), \tag{2}$$

where

$$\omega(u(t),t) = [\omega_1(u_1(t),t),...,\omega_L(u_L(t),t)]. \tag{3}$$

Let us consider a single input case for hysteresis nonlinearities modeled as [21,41]

$$\omega(u) = \mu_1 u + \mu_2 \zeta, \tag{4}$$

where μ_1 and μ_2 are non-zero constants with $\mu_1\mu_2 > 0$, and ζ satisfies

$$\dot{\zeta} = \dot{u} - \omega|\dot{u}||\zeta|^{n-1}\zeta - \beta\dot{u}|\zeta|^n = \dot{u}\psi(\zeta,\dot{u}), \tag{5}$$

where $n \geq 1$, $\omega > |\beta|$ and $\psi(\zeta,\dot{u}) = 1 - \omega\text{sign}(\dot{u})|\zeta|^{n-1} - \beta|\zeta|^n$. From [21,41], one has $|\zeta(t)| \leq \sqrt[n]{1/(\omega+\beta)}$. The solution of Equation (4) is depicted in Figure 1, where the hysteresis parameters are set as $\omega = 1$, $n = 1$, $\mu_1 = 4.5$, $\beta = 0$, $\mu_2 = 4$, and $u = 10\sin(5t)$. As shown in Figure 1, the nominal input dynamics preceded by hysteresis phenomena are nonlinear when compared to the linear case wherein $v = u$.

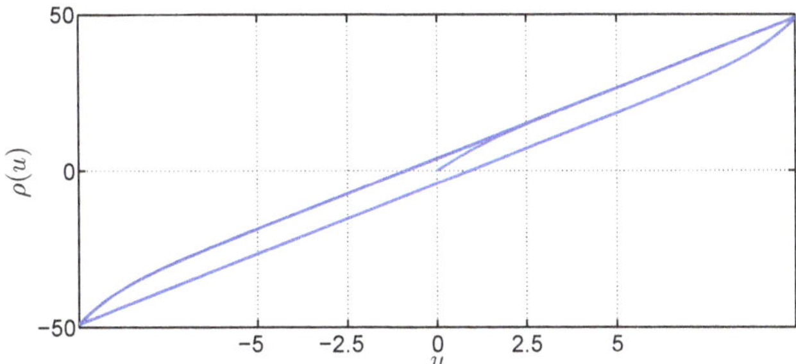

Figure 1. Hysteresis nonlinearities simulated using (4).

The control objective in this paper is to construct a two-adaptive-laws-based control scheme for the robotic manipulator (1) with unknown input Bouc–Wen hysteresis (2) so that outputs of the robotic manipulator $q(t)$ track to desired trajectories, that is,

$$\lim_{t\to\infty}(q(t) - q_d(t)) = 0, \tag{6}$$

$$\lim_{t\to\infty}(\dot{q}(t) - \dot{q}_d(t)) = 0. \tag{7}$$

To summarize the design purpose, we give the closed-loop system after applying the control scheme to the robotic manipulators in Figure 2.

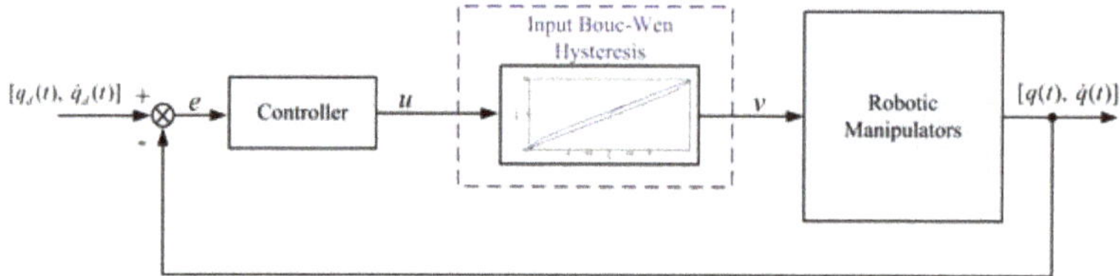

Figure 2. Control diagram for robotic manipulators with unknown input Bouc–Wen hysteresis.

3. Trajectory Tracking Design for Robotic Manipulators with Unknown Input Hysteresis

In this section, we specify the control method, control design, and the main result for robotic manipulators with unknown input hysteresis. We show that trajectory tracking control is ensured using the proposed adaptive control in the sense of Lyapunov theory.

3.1. Control Method

In this subsection, we review the dynamical gain-based approach [42], which will be combined with the adaptive control technique to handle unknown coefficients caused by input hysteresis.

Here, the dynamical gain is given as [42]

$$\mathcal{N}(\chi) = \omega e^{\chi^2},\tag{8}$$

where χ is a real variable. Recalling the result in [42], one has the following result:

Lemma 1. *Let functions $V(t)$ and $\chi(t)$ smooth over $[0, t_s)$ with $V(t) \geq 0$ and $\chi(0)$ bounded. Moreover, $\chi(t)$ is a monotonic function. The dynamic loop gain function $\mathcal{N}$ is as (8). If one has*

$$V(t) \leq \beta + \int_{t_0}^{t} \dot{\chi}(\omega)e^{\Im\omega - \Im t}d\omega$$

$$- \int_{t_0}^{t} g_M \mathcal{N}(\chi(\omega))\dot{\chi}(\omega)e^{\Im\omega - \Im t}d\omega,\tag{9}$$

where β is a bounded variable and $\Im$ and g_M are positive constants, then the boundedness of $V(t)$ and $\chi(t)$ are derived over $[0, t_s)$.

Lemma 2 (Barbalat's Lemma [3]). *Let a function $f(t) \in \mathcal{C}^1(a, \infty)$ and $\lim_{t \to \infty} f(t) = a$, where $a < \infty$. If f' is uniformly continuous, then $\lim_{t \to \infty} f'(t) = 0$.*

In what follows, we show how to use dynamical gain (8) to handle unknown input hysteresis in robotic manipulators and how to use one parameter to adaptively tune control coefficients for multiple inputs.

3.2. Controller Design

In this subsection, we show the control design to handle input hysteresis in robotic manipulators. Recalling the work in [3], we define

$$\tilde{q} = q - q_d,\tag{10}$$

$$\dot{q}_r = \dot{q}_d - K_r\tilde{q},\tag{11}$$

$$e = \dot{q} - \dot{q}_r,\tag{12}$$

where $K_r \in \mathbb{R}^{L \times L} > 0$ is a positive-definite matrix.

Now, a controller for robotic manipulators to reject input hysteresis is given as

$$u = \mathcal{N}(\chi(t))u_{\mathcal{N}}, \tag{13}$$

with

$$u_{\mathcal{N}} = -\frac{1}{2}||\Psi||_2 \hat{\lambda}e - (K_e + \frac{1}{2}I)e, \tag{14}$$

where $\hat{\lambda}$ is an estimate of λ to be detailed later, K_e is a positive-definite matrix, and I denotes an identity matrix with the dimension of $\mathbb{R}^{L \times L}$. The adaptive laws for (13) and (14) are given as

$$\dot{\chi} = \omega e^T u_{\mathcal{N}}, \tag{15}$$

$$\dot{\hat{\lambda}} = \frac{1}{2}\Phi ||\Psi(q, \dot{q}, \dot{q}_r, \ddot{q}_r)||_2 e^T e - \Phi_1 \hat{\lambda}, \tag{16}$$

where ω, Φ, and Φ_1 denote positive constants and $|| \cdot ||_2$ denotes a norm operator. The initial values of $\lambda(t)$ and $\chi(t)$ are set to be non-negative, i.e., $\lambda(0) \geq 0$ and $\chi(0) \geq 0$. To summarize the design purpose, we show the designed control scheme in Figure 3.

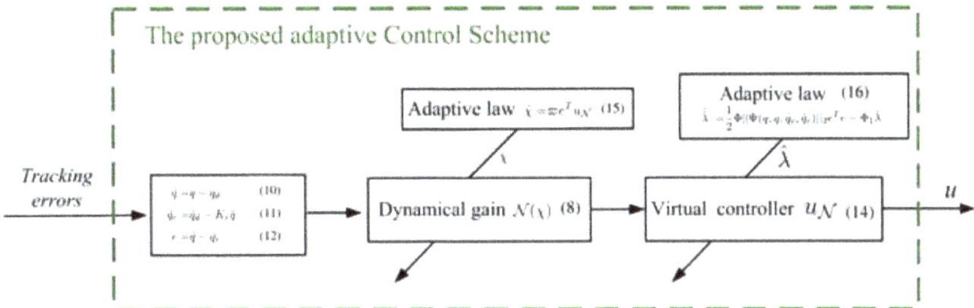

Figure 3. Proposed adaptive control scheme using a dynamical gain-based method.

3.3. Stability Analysis

Based on the control design in the previous subsection, we use Lyapunov theory to analyze the stability of the proposed adaptive control with a focus on handling unknown input Bouc–Wen hysteresis. Our main result is summarized as follows.

Theorem 1. *Supposing that the robotic manipulators are modeled as (1) with input hysteresis as (4) and (5), the controller is given as (13), and adaptive mechanisms are as (15) and (16), the asymptotic tracking performance in terms of $q(t)$ and $\dot{q}(t)$ in (1) is guaranteed such that $q(t) \to q_d(t)$ and $\dot{q}(t) \to \dot{q}_d(t)$ as $t \to \infty$.*

Proof. Substituting (13) into (1) leads to

$$D(q)\dot{e} + H(q, \dot{q})e = [\omega(u, t) - I]u_{\mathcal{N}} + u_{\mathcal{N}} - \Psi\theta, \tag{17}$$

where Ψ is a regression matrix and θ is a constant vector with an appropriate dimension. Define a function

$$V = \frac{1}{2}e^T D(q)e + \frac{1}{2}\tilde{\lambda}^2 \Phi^{-1}, \tag{18}$$

where $\tilde{\lambda}$ is defined as

$$\tilde{\lambda} = \hat{\lambda} - \lambda, \tag{19}$$

with λ being defined later and $D(q)$ being a positive definite matrix following [2]. From (18), one has

$$\dot{V} = u_\mathcal{N} + e^T \left[\omega(u,t) - u_\mathcal{N} \right] + e^T \Psi \theta + \hat{\lambda} \tilde{\lambda} \Phi^{-1}$$

$$= u_\mathcal{N} + e^T \left[\omega(u,t) - u_\mathcal{N} \right] + \hat{\lambda} \tilde{\lambda} \Phi^{-1} + \frac{1}{2} + \frac{1}{2} ||\Psi||_2 \lambda e^T e, \tag{20}$$

where $\lambda = ||\theta||_F$. Following [2,4], one has that $\dot{D}(q) - 2H(q,\dot{q})$ is a skew-symmetric matrix. Substituting (14) and (16) into (20) yields

$$\dot{V} \leq - e^T K_e e - \frac{1}{2} e^T e + e^T \left[\omega(u,t) - u_\mathcal{N} \right] + \frac{1}{2}$$

$$+ \hat{\lambda} \tilde{\lambda} \Phi^{-1} - \frac{1}{2} ||\Psi||_2 e^T e \tilde{\lambda}$$

$$= - e^T K_e e - \frac{1}{2} e^T e + e^T \left[\omega(u,t) - u_\mathcal{N} \right] + \frac{1}{2} - \Phi_1 \hat{\lambda} \tilde{\lambda} \Phi^{-1}, \tag{21}$$

where the first inequality is derived after using Young's inequality. Now, $-\hat{\lambda}\tilde{\lambda}$ in the right-hand side of (21) is changed into

$$-\hat{\lambda}\tilde{\lambda} = -(\tilde{\lambda} + \lambda)\tilde{\lambda}$$

$$= -\tilde{\lambda}\tilde{\lambda} + \tilde{\lambda}\lambda$$

$$\leq -\tilde{\lambda}\tilde{\lambda} + \frac{1}{2}\tilde{\lambda}^2 + \frac{1}{2}\lambda^2$$

$$= -(\frac{1}{2}\tilde{\lambda}^2 - \frac{1}{2}\lambda^2), \tag{22}$$

where both the result in (19) and Young's inequality are used. From (15) and (22), (21) is further changed into

$$\dot{V} \leq - e^T K_e e - \frac{1}{2} e^T e + e^T \left[\omega(u,t) - u_\mathcal{N} \right] + \frac{1}{2}$$

$$- \Phi_1 \Phi^{-1} (\frac{1}{2}\tilde{\lambda}^2 - \frac{1}{2}\lambda^2)$$

$$\leq - e^T K_e e - \Phi_1 \Phi^{-1} \frac{1}{2}\tilde{\lambda}^2 + e^T \left[(Diag(\mu_1)\mathcal{N}(\chi(t))u_\mathcal{N} \right.$$

$$\left. + Diag(\mu_2)Vec(\zeta)) - u_\mathcal{N} \right] + \frac{1}{2} - \frac{1}{2} e^T e + \Phi_1 \Phi^{-1} \frac{1}{2}\lambda^2$$

$$\leq - e^T K_e e - \Phi_1 \Phi^{-1} \frac{1}{2}\tilde{\lambda}^2 + e^T (\mathcal{N}(\chi(t))\mu_{\min} - 1)u_\mathcal{N}$$

$$+ \frac{1}{2}\mu_{\max} + \Phi_1 \Phi^{-1} \frac{1}{2}\lambda^2 + \frac{1}{2}$$

$$\leq - \Im V(t) + e^T (\mathcal{N}(\chi(t))\mu_{\min} - 1)u_\mathcal{N} + \beta_0$$

$$\leq - \Im V(t) + (\mathcal{N}(\chi(t))\mu_{\min} - 1)\dot{\chi}(t)\frac{1}{\omega} + \beta_0, \tag{23}$$

where

$$\Im = \min\left\{\frac{2\delta_{\min}\{K_e\}}{\delta_{\min}(D(q))}, \Phi_1\right\}, \tag{24}$$

$$\beta_0 = \frac{1}{2}\mu_{\max} + \Phi_1\Phi^{-1}\frac{1}{2}\lambda^2 + \frac{1}{2}, \tag{25}$$

$$Diag(\mu_1) = Diag[\mu_{1,1}, \mu_{1,2}, ..., \mu_{1,L}], \tag{26}$$

$$Diag(\mu_2) = Diag[\mu_{2,1}, \mu_{2,2}, ..., \mu_{2,L}], \tag{27}$$

$$Vec(\zeta) = [\zeta_1, \zeta_2, \cdots, \zeta_L], \tag{28}$$

$$\mu_{\min} = \min_{i,j=1,2,\cdots,L}(\mu_{i,j}), \tag{29}$$

$$\mu_{\max} = ||Diag(\mu_2)Vec(\zeta)||_2, \tag{30}$$

$$Vec(\zeta) = [\zeta_1, \zeta_2, \cdots, \zeta_L]^T. \tag{31}$$

It is clear that $\Im$, $\mu_{\min}$, and β_0 are positive constants. $\quad\square$

Remark 1. *Here, (23) specifies how to transform the unknown input Bouc–Wen hysteresis control problem into a problem of handling an unknown control coefficient $\mu_{\min}$ and an unknown variable β_0, where $\mu_{\min}$ is given in (29) and β_0 is given in (25). We use the dynamical loop gain (8) and the designed control scheme (14) to make sure that the multiplication in (13) is non-positive. Based on the non-positiveness of $\mathcal{N}(\chi(t))u_{\mathcal{N}}$, one now finds a upper bound governed by the minimum (29). Please note that even though robotic manipulators are modeled as multiple-input and multiple-output systems, one only needs to handle two scalars, $\mu_{\min}$ and β_0, in (23). This further prompts our adaptive method that uses two adaptive laws to achieve asymptotic control.*

From (23), one has

$$\begin{aligned}
V(t) \leq &-V(0)e^{-\Im t} + \frac{\Im}{\beta_0} + \frac{1}{\varpi}\int_{t_0}^t \dot{\chi}(\omega)e^{\Im\omega-\Im t}d\omega \\
&-\frac{1}{\varpi}\int_{t_0}^t \mu_{\min}\mathcal{N}(\chi(\omega))\dot{\chi}(\omega)e^{\Im\omega-\Im t}d\omega \\
\triangleq &\,\beta + \frac{1}{\varpi}\int_{t_0}^t \dot{\chi}(\omega)e^{\Im\omega-\Im t}d\omega \\
&-\frac{1}{\varpi}\int_{t_0}^t \mu_{\min}\mathcal{N}(\chi(\omega))\dot{\chi}(\omega)e^{\Im\omega-\Im t}d\omega,
\end{aligned} \tag{32}$$

where

$$\beta = -V(0)e^{-\Im t} + \frac{\Im}{\beta_0}. \tag{33}$$

Considering that $V(0)$ is predetermined to be bounded and $\Im$ and β_0 are bounded, one obtains that β is also bounded.

Now, we obtain that (32) is structurally the same as (9). Therefore, the result in *Lemma 1* will hold for (32). That is, from the result in *Lemma 1*, one obtains the boundedness of $V(t)$ and $\chi(t)$ [37]. As an immediate result from (14) and (15), one has

$$\begin{aligned}
\chi(t) - \chi(0) &= \int_0^t \varpi e^T(\tau)u_{\mathcal{N}}(\tau)d\tau \\
&\geq \int_0^t \varpi e^T(\tau)e(\tau)d\tau.
\end{aligned} \tag{34}$$

Note that the boundedness of $\chi(t)$ and $\chi(0)$ has been ensured and ω is a predetermined constant. It is clear that $\int_0^t \omega e^T(\tau)e(\tau)d\tau$ exists and is finite. From *Lemma 2 (Barbalat's Lemma)*, one has

$$\lim_{t\to\infty} e^T e = 0, \tag{35}$$

so that

$$\lim_{t\to\infty} e(t) = 0. \tag{36}$$

Therefore, the convergence in (6) and (7) results. Thus, the proof is completed.

Remark 2. *In Theorem 1, we have proven that even though multiple inputs coexist in the considered robotic manipulators, as shown in (3), the control objective is achieved with computational efficiency using two parameters to be tuned online. In particular, one is in (15) and is responsible for handling unknown input coefficients from hysteresis, and the other one is in (16) and is responsible for addressing the regression matrix from robotic manipulators. This two-parameter adaptive control scheme is feasible due to our unique control design as in (14) and the dynamical gain as in (8). Furthermore, we employ the adaptive control technique to achieve stability for the trajectory tracking control of robotic manipulators with unknown input Bouc–Wen hysteresis.*

4. Simulation Example

A two-link articulated robotic manipulator is used for the simulation, which follows the work of [3]. The proposed method is employed to testify to the validity of the proposed control scheme. The manipulator is simulated to move in a horizontal plane and is described as in [3]:

$$\begin{bmatrix} D_{11} & D_{12} \\ D_{21} & D_{22} \end{bmatrix} \begin{bmatrix} \ddot{q}_1 \\ \ddot{q}_2 \end{bmatrix} + \begin{bmatrix} H_{11} & H_{12} \\ H_{21} & H_{22} \end{bmatrix} \begin{bmatrix} \dot{q}_1 \\ \dot{q}_2 \end{bmatrix} = \begin{bmatrix} v_1 \\ v_2 \end{bmatrix}, \tag{37}$$

where

$$D_{11} = \theta_1 + 2\theta_3 \cos(q_2),$$

$$D_{12} = H_{21} = \theta_2 + \theta_3 \cos(q_2) + \theta_4 \sin(q_2),$$

$$D_{22} = \theta_2,$$

$$H_{11} = -h\dot{q}_2, \quad H_{12} = -h(\dot{q}_1 + \dot{q}_2),$$

$$H_{21} = h\dot{q}_1, \quad H_{22} = 0,$$

and

$$h = \theta_3 \sin(q_2) - \theta_4 \cos(q_2),$$

with the physical parameters being $\theta = [\theta_1, \theta_2, \theta_3, \theta_4]^T$. Here, it is noted that the considered system (37) has two inputs and two outputs. As for the system inputs, we specify the input Bouc–Wen hysteresis for each torque as in (4) and (5) with hysteresis parameters $\mu_1 = 4.5, \mu_2 = 4, \omega = 1, \beta = 0$, and $n = 1$. Note that the parameters of hysteresis nonlinearity for each torque can be different according to our result in Section 3. Here, we choose the same hysteresis parameters for simplification. The initial states for the robotic manipulators are also chosen randomly. Note that we need two adaptive laws to implement our method. Here, we set the initials of these two adaptive laws in (15) and (16) as zeroes. That is, $\hat{\lambda}(0) = 0$ and $\chi(0) = 0$. Note that the physical parameters to be estimated are vectorized as $\theta = [\theta_1, \theta_2, \theta_3, \theta_4]^T$. Here, we we use the adaptive law (16) to estimate the scalar $\lambda = ||\theta||_F$, not the vector θ. Therefore, the number of estimators drops significantly to only one when compared to the traditional adaptive method. As a result, the computational efficiency is ensured by our method.

In what follows, we give two scenarios that frequently happen in the motion control of robotic manipulators.

4.1. Fixed-Point Control Using the Proposed Adaptive Control

In this scenario, we predetermine the desired trajectory as the predetermined points $q_d = [q_{d1}, q_{d2}]^T = [0.2, 0.5]^T$ and $\dot{q}_d = [\dot{q}_{d1}, \dot{q}_{d2}]^T = [0, 0]^T$. Simulation results are given in Figures 4–9. From the observation of Figures 4–7, the adaptive variables including χ, $\mathcal{N}(\chi)$, u, and $\hat{\lambda}$ are bounded under unknown input hysteresis and the proposed control method. The outputs of the considered robotic manipulators q and $\dot{q}$, as well as the predetermined ones q_d and $\dot{q}_d$, are shown in Figures 8 and 9, where outputs $q_1(t)$ and $q_2(t)$ are, respectively, driven to the predetermined points 0.2 and 0.5 in the presence of the proposed control, while the velocities $\dot{q}_1(t)$ and $\dot{q}_2(t)$ are regularized to zeroes. Therefore, it is clear that the proposed method is effective in handling input hysteresis in robotic manipulators for the fixed-point control.

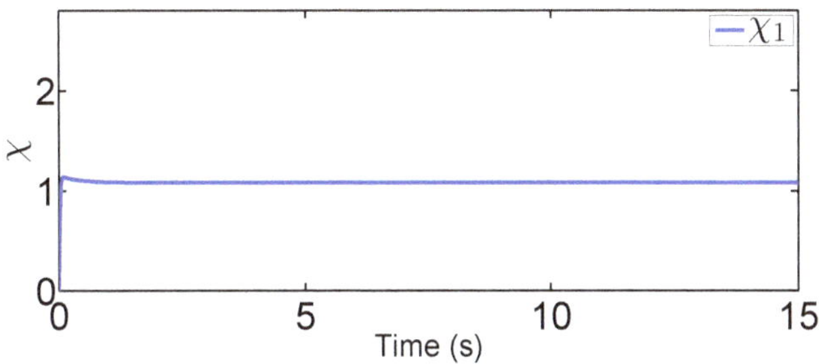

Figure 4. Adaptive law χ for fixed-point control.

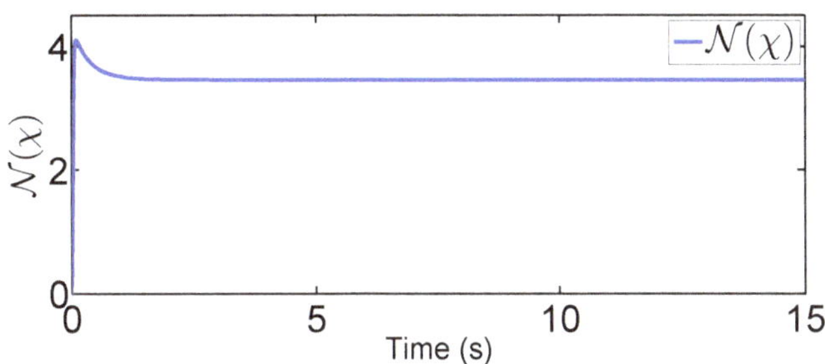

Figure 5. Dynamic loop gain function for fixed-point control.

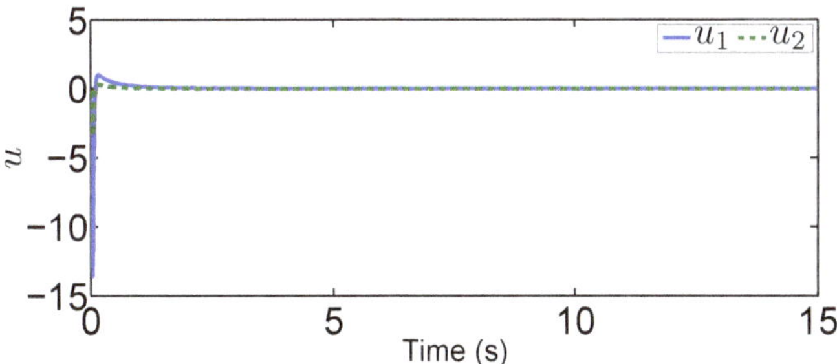

Figure 6. Input signal u for fixed-point control.

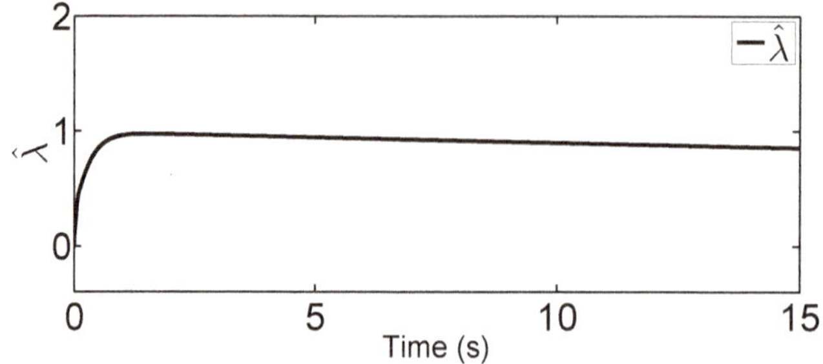

Figure 7. Adaptive law $\hat{\lambda}$ for fixed-point control.

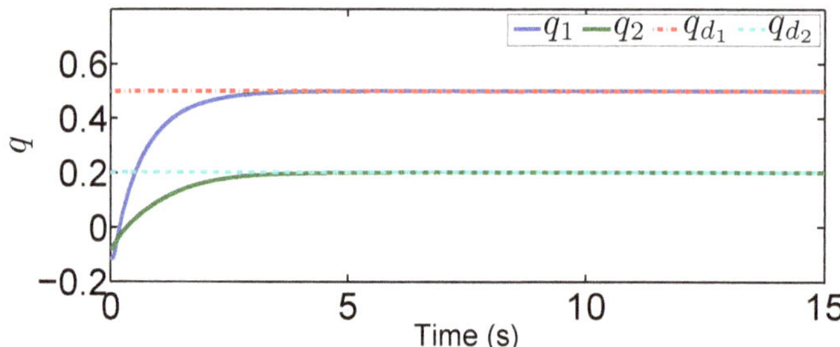

Figure 8. Output q for fixed-point control.

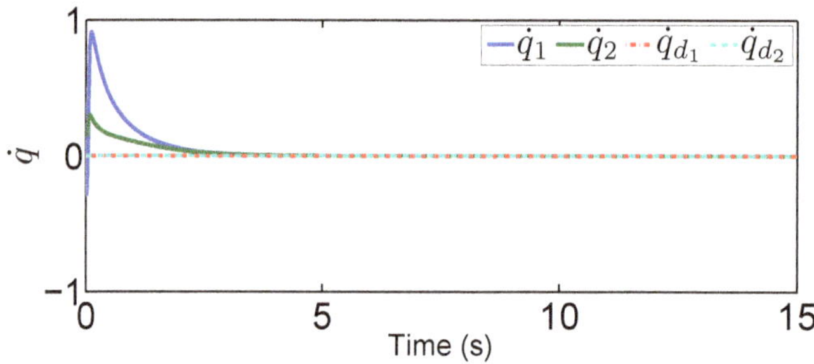

Figure 9. Output $\dot{q}$ for fixed-point control.

4.2. Tracking Control Using the Proposed Adaptive Control

In this tracking control scenario, we set the desired trajectory to be a sine wave. Simulation results for this scenario including the adaptive signals χ, $\mathcal{N}(\chi)$, $\hat{\lambda}$, and control signal u, are given in Figures 10–15. In particular, the signal of χ and its dynamical gain $\mathcal{N}(\chi)$ are given in Figures 10 and 11. The control signal u is given in Figure 12. The adaptive law of $\hat{\lambda}$ is given in Figure 13. The results in Figures 10–13 show that our method is effective in ensuring all the signals in the closed-loop robotic manipulator are bounded. Finally, the outputs q and $\dot{q}$ are provided in Figures 14 and 15, where the proposed method drives the outputs of robotic manipulators to converge to the desired trajectories. This guarantees the effectiveness of the proposed method in achieving tracking control in the presence of unknown input hysteresis.

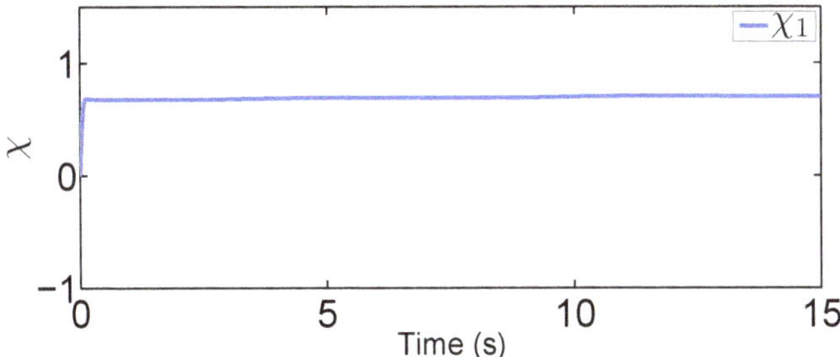

Figure 10. Adaptive law χ for tracking control.

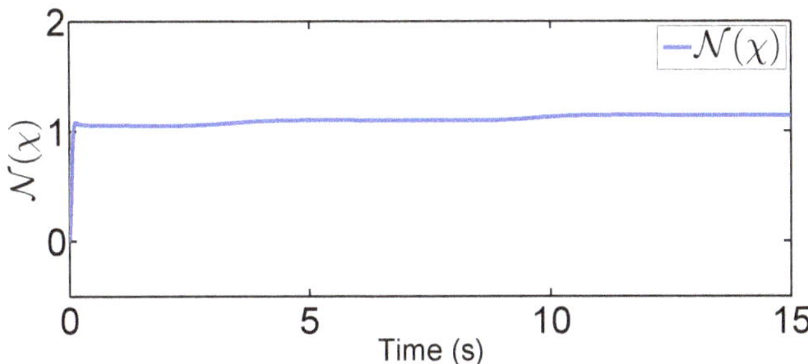

Figure 11. Dynamic loop gain function for tracking control.

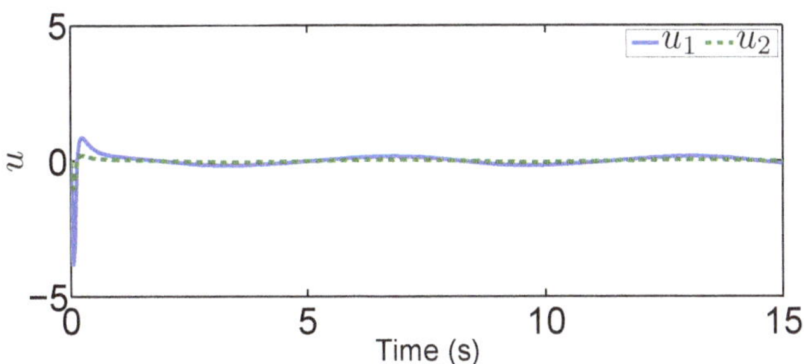

Figure 12. Input signal u for tracking control.

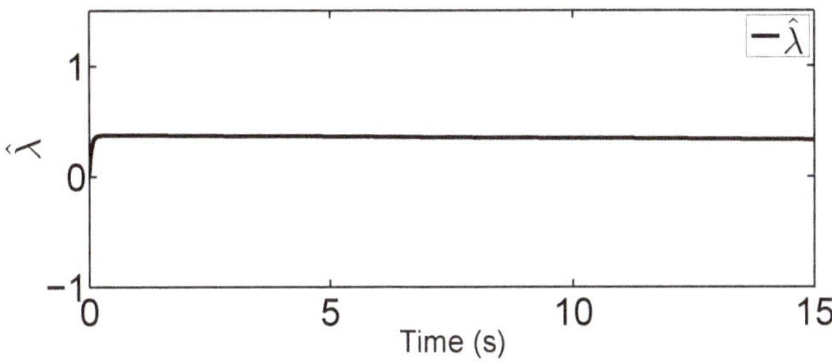

Figure 13. Adaptive law $\hat{\lambda}$ for tracking control.

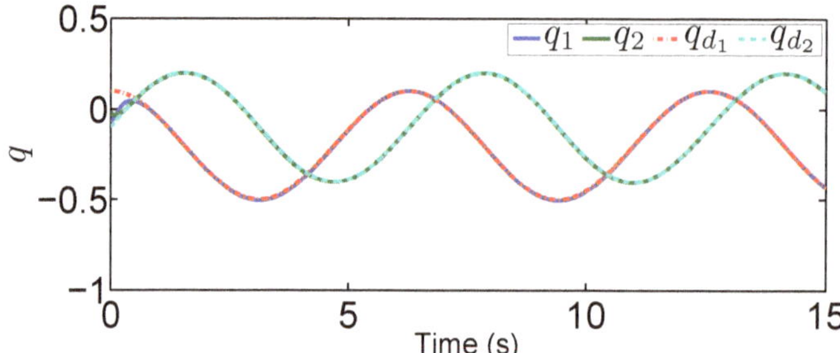

Figure 14. Output q for tracking control.

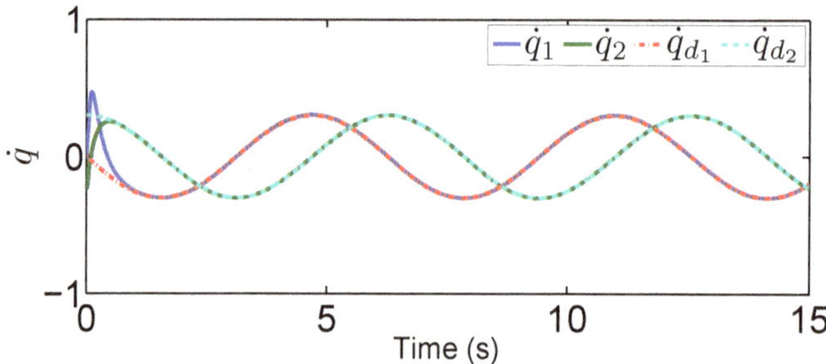

Figure 15. Output $\dot{q}$ for tracking control.

Moreover, we give the tracking performance under a traditional controller without compensating the hysteresis nonlinearities. For the comparison, we consider the same two-link robotic manipulator as in the previous case. To be specific, a proportional plus derivative controller is applied with $v = -\frac{1}{2}(\dot{q} - \dot{q}_d) - \frac{5}{2}(q - q_d)$. The tracking performance in the presence of the traditional controller is given in Figures 16 and 17. From Figures 14–17, it is clear that our method provides a better tracking performance compared to the traditional controller.

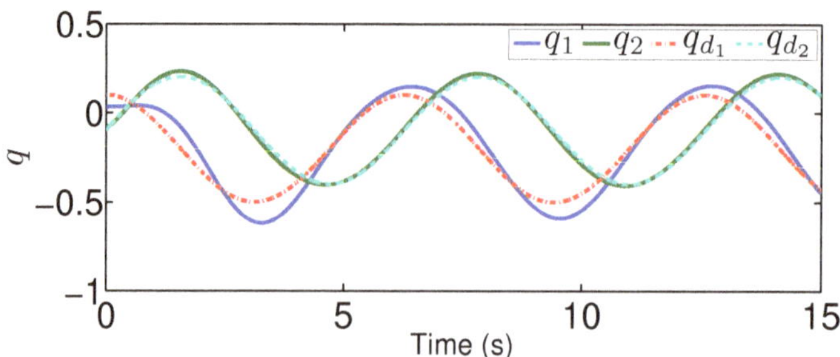

Figure 16. Output q for tracking control under the traditional controller.

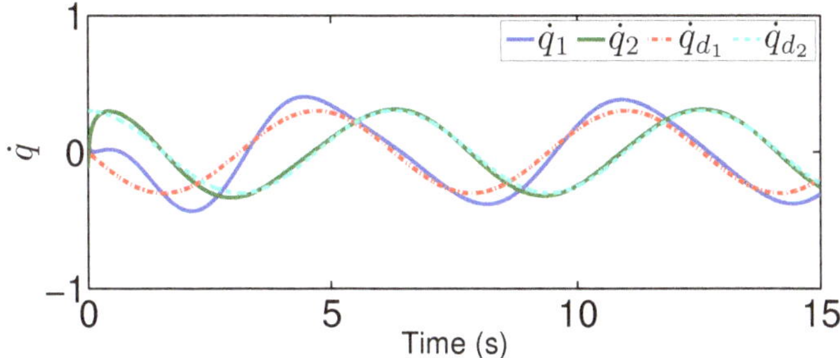

Figure 17. Output $\dot{q}$ for tracking control under the traditional controller.

5. Conclusions

In this paper, the problem of input hysteresis is addressed for robotic manipulators. We utilize the adaptive control technique and a dynamical gain-based approach to handle input hysteresis. We use two adaptive parameters to address input hysteresis in robotic manipulators so that computational efficiency is ensured for real-time implementation. Therefore, the proposed adaptive method may be feasible for the purpose of applications. Moreover, we drive the outputs of robotic manipulators to the desired trajectories with zero errors, which guarantees a high level of control quality for robotic manipulators even in presence of unknown input hysteresis. We adopt Lyapunov theory to validate the stability of our method and to prove that all the states and adaptive variables in the closed-loop systems are bounded. In addition, we provide a numerical example including fixed-point and trajectory controls so that the validity of our method is ensured. Future works may extend the proposed method and combine it with advanced learning methods such as those in [43–49].

Author Contributions: K.X. and W.L. conceived of the original idea of the paper. K.X. and Y.L. performed the experiments. K.X., Y.L., and W.L. wrote the paper.

Acknowledgments: This work was supported by the National Natural Science Foundation of China under Grant 61703113.

Conflicts of Interest: The authors declare no conflict of interest.

References

1. Siciliano, B.; Khatib, O. *Springer Handbook of Robotics*; Springer: Berlin/Heidelberg, Germany, 2008.
2. Lewis, F.L.; Dawson, D.M.; Abdallah, C.T. *Robot Manipulator Control: Theory and Practice*; Marcel Dekker: New York, NY, USA, 2003.
3. Slotine, J.J.E.; Li, W. *Applied Nonlinear Control*; Prentice-Hall: Englewood Cliffs, NJ, USA, 1991.
4. Spong, M.W.; Vidyasagar, M. *Robot Dynamics and Control*; John Wiley & Sons: New York, NY, USA, 2008.
5. Tran, D.T.; Truong, H.V.A.; Ahn, K.K. Adaptive Backstepping Sliding Mode Control Based RBFNN for a Hydraulic Manipulator Including Actuator Dynamics. *Appl. Sci.* **2019**, *9*, 1265. [CrossRef]
6. Li, Z.; Yang, Z.; Xie, S. Computing Resource Trading for Edge-Cloud-assisted Internet of Things. *IEEE Trans. Ind. Inform.* **2019**, *15*, 3661–3669, doi:10.1109/TII.2019.2897364. [CrossRef]
7. Liu, Y.; Yang, C.; Jiang, L.; Xie, S.; Zhang, Y. Intelligent Edge Computing for IoT-Based Energy Management in Smart Cities. *IEEE Netw.* **2019**, *33*, 111–117, doi:10.1109/MNET.2019.1800254. [CrossRef]
8. Yan, J.; Ban, H.; Luo, X.; Zhao, H.; Guan, X. Joint Localization and Tracking Design for AUV With Asynchronous Clocks and State Disturbances. *IEEE Trans. Veh. Technol.* **2019**, *68*, 4707–4720, doi:10.1109/TVT.2019.2903212. [CrossRef]
9. Vo, A.T.; Kang, H.J. An Adaptive Neural Non-Singular Fast-Terminal Sliding-Mode Control for Industrial Robotic Manipulators. *Appl. Sci.* **2018**, *8*, 2562. [CrossRef]
10. Zhou, J.; Zhang, C.; Wen, C. Robust adaptive output control of uncertain nonlinear plants with unknown backlash nonlinearity. *IEEE Trans. Autom. Control* **2007**, *52*, 503–509. [CrossRef]

11. Fu, M.; Xie, L. The sector bound approach to quantized feedback control. *IEEE Trans. Autom. Control* **2005**, *50*, 1698–1711.
12. Fu, M.; Xie, L. Finite-Level Quantized Feedback Control for Linear Systems. *IEEE Trans. Autom. Control* **2009**, *54*, 1165–1170, doi:10.1109/TAC.2009.2017815. [CrossRef]
13. Hayakawa, T.; Ishii, H.; Tsumura, K. Adaptive quantized control for nonlinear uncertain systems. *Syst. Control Lett.* **2009**, *58*, 625–632. [CrossRef]
14. Zhou, J.; Wen, C.; Yang, G. Adaptive backstepping stabilization of nonlinear uncertain systems with quantized input signal. *IEEE Trans. Autom. Control* **2014**, *59*, 460–464. [CrossRef]
15. Chen, C.; Wen, C.; Liu, Z.; Xie, K.; Zhang, Y.; Chen, C.L.P. Adaptive Consensus of Nonlinear Multi-Agent Systems with Non-Identical Partially Unknown Control Directions and Bounded Modelling Errors. *IEEE Trans. Autom. Control* **2017**, *62*, 4654–4659. [CrossRef]
16. Chen, C.; Xie, K.; Lewis, F.L.; Xie, S.; Davoudi, A. Fully Distributed Resilience for Adaptive Exponential Synchronization of Heterogeneous Multi-Agent Systems Against Actuator Faults. *IEEE Trans. Autom. Control* **2018**, doi:10.1109/TAC.2018.2881148. [CrossRef]
17. Xie, K.; Chen, C.; Lewis, F.L.; Xie, S. Adaptive Asymptotic Neural Network Control of Nonlinear Systems With Unknown Actuator Quantization. *IEEE Trans. Neural Netw. Learn. Syst.* **2018**, *29*, 6303–6312, doi:10.1109/TNNLS.2018.2828315. [CrossRef]
18. Chen, C.; Lewis, F.L.; Xie, S.; Modares, H.; Liu, Z.; Zuo, S.; Davoudi, A. Resilient adaptive and H_∞ controls of multi-agent systems under sensor and actuator faults. *Automatica* **2019**, *102*, 19–26. [CrossRef]
19. Cao, K.; Li, R. Modeling of Rate-Independent and Symmetric Hysteresis Based on Madelung's Rules. *Sensors* **2019**, *19*, 352. [CrossRef] [PubMed]
20. Wen, C.; Zhou, J. Decentralized adaptive stabilization in the presence of unknown backlash-like hysteresis. *Automatica* **2007**, *43*, 426–440. [CrossRef]
21. Zhou, J.; Wen, C.Y.; Li, T.S. Adaptive output feedback control of uncertain nonlinear systems with hysteresis nonlinearity. *IEEE Trans. Autom. Control* **2012**, *57*, 2627–2633. [CrossRef]
22. Lin, J.; Chiang, M. Tracking Control of a Magnetic Shape Memory Actuator Using an Inverse Preisach Model with Modified Fuzzy Sliding Mode Control. *Sensors* **2016**, *16*, 1368. [CrossRef] [PubMed]
23. Yan, J.; Wan, Y.; Luo, X.; Chen, C.; Hua, C.; Guan, X. Formation Control of Teleoperating Cyber-Physical System With Time Delay and Actuator Saturation. *IEEE Trans. Control Syst. Technol.* **2018**, *26*, 1458–1467, doi:10.1109/TCST.2017.2709266. [CrossRef]
24. Bai, X.; Cai, F.; Chen, P. Resistor-capacitor (RC) operator-based hysteresis model for magnetorheological (MR) dampers. *Mech. Syst. Signal Process.* **2019**, *117*, 157–169. [CrossRef]
25. Chen, P.; Bai, X.; Qian, L.; Choi, S. An Approach for Hysteresis Modeling Based on Shape Function and Memory Mechanism. *IEEE/ASME Trans. Mech.* **2018**, *23*, 1270–1278, doi:10.1109/TMECH.2018.2833459. [CrossRef]
26. Nussbaum, R.D. Some remarks on a conjecture in parameter adaptive control. *Syst. Control Lett.* **1983**, *3*, 243–246. [CrossRef]
27. Mårtensson, B. Remarks on adaptive stabilization of first order non-linear systems. *Syst. Control Lett.* **1990**, *14*, 1–7. [CrossRef]
28. Ge, S.S.; Wang, J. Robust adaptive neural control for a class of perturbed strict feedback nonlinear systems. *IEEE Trans. Neural Netw.* **2002**, *13*, 1409–1419. [CrossRef] [PubMed]
29. Ding, Z. Adaptive control of non-linear systems with unknown virtual control coefficients. *Int. J. Adapt. Control Signal Process.* **2000**, *14*, 505–517. [CrossRef]
30. Ye, X.; Jiang, J. Adaptive nonlinear design without a priori knowledge of control directions. *IEEE Trans. Autom. Control* **1998**, *43*, 1617–1621.
31. Zhang, Y.; Wen, C.Y.; Soh, Y.C. Adaptive backstepping control design for systems with unknown high-frequency gain. *IEEE Trans. Autom. Control* **2000**, *45*, 2350–2354. [CrossRef]
32. Ge, S.S.; Wang, J. Robust adaptive tracking for time-varying uncertain nonlinear systems with unknown control coefficients. *IEEE Trans. Autom. Control* **2003**, *48*, 1463–1469.
33. Kim, B.; Park, B.S. Robust Control for the Segway with Unknown Control Coefficient and Model Uncertainties. *Sensors* **2016**, *16*, 1000. [CrossRef]
34. Ye, X. Decentralized adaptive regulation with unknown high-frequency-gain signs. *IEEE Trans. Autom. Control* **1999**, *44*, 2072–2076.

35. Ge, S.S.; Hong, F.; Lee, T.H. Adaptive neural control of nonlinear time-delay systems with unknown virtual control coefficients. *IEEE Trans. Syst. Man Cybern. B Cybern.* **2004**, *34*, 499–516. [CrossRef] [PubMed]
36. Ye, X. Decentralized adaptive stabilization of large-scale nonlinear time-delay systems with unknown high-frequency-gain signs. *IEEE Trans. Autom. Control* **2011**, *56*, 1473–1478. [CrossRef]
37. Chen, W.S.; Li, X.B.; Ren, W.; Wen, C.Y. Adaptive Consensus of Multi-Agent Systems With Unknown Identical Control Directions Based on A Novel Nussbaum-Type Function. *IEEE Trans. Autom. Control* **2014**, *59*, 1887–1892. [CrossRef]
38. Chen, C.; Liu, Z.; Zhang, Y.; Chen, C.L.P.; Xie, S. Saturated Nussbaum Function based Approach for Robotic Systems with Unknown Actuator Nonlinearities. *IEEE Trans. Cybern.* **2016**, *46*, 2311–2322. [CrossRef]
39. Yan, J.; Li, X.; Luo, X.; Guan, X. Virtual-Lattice Based Intrusion Detection Algorithm over Actuator-Assisted Underwater Wireless Sensor Networks. *Sensors* **2017**, *17*, 1168. [CrossRef] [PubMed]
40. Chen, C.; Liu, Z.; Xie, K.; Zhang, Y.; Chen, C.P. Asymptotic adaptive control of nonlinear systems with elimination of overparametrization in a Nussbaum-like design. *Automatica* **2018**, *98*, 277–284, doi:10.1016/j.automatica.2018.09.034. [CrossRef]
41. Ikhouane, F.; MañOsa, V.; Rodellar, J. Adaptive control of a hysteretic structural system. *Automatica* **2005**, *41*, 225–231. [CrossRef]
42. Chen, C.; Wen, C.; Liu, Z.; Xie, K.; Zhang, Y.; Chen, C.P. Adaptive asymptotic control of multivariable systems based on a one-parameter estimation approach. *Automatica* **2017**, *83*, 124–132. [CrossRef]
43. Xie, S.; Yang, L.; Yang, J.M.; Zhou, G.; Xiang, Y. Time-Frequency Approach to Underdetermined Blind Source Separation. *IEEE Trans. Neural Netw. Learn. Syst.* **2012**, *23*, 306–316, doi:10.1109/TNNLS.2011.2177475. [CrossRef]
44. Zhou, G.; Cichocki, A.; Zhang, Y.; Mandic, D.P. Group Component Analysis for Multiblock Data: Common and Individual Feature Extraction. *IEEE Trans. Neural Netw. Learn. Syst.* **2016**, *27*, 2426–2439, doi:10.1109/TNNLS.2015.2487364. [CrossRef]
45. He, Z.; Cichocki, A.; Xie, S.; Choi, K. Detecting the Number of Clusters in n-Way Probabilistic Clustering. *IEEE Trans. Pattern Anal. Mach. Intell.* **2010**, *32*, 2006–2021, doi:10.1109/TPAMI.2010.15. [CrossRef] [PubMed]
46. He, Z.; Xie, S.; Zdunek, R.; Zhou, G.; Cichocki, A. Symmetric Nonnegative Matrix Factorization: Algorithms and Applications to Probabilistic Clustering. *IEEE Trans. Neural Netw.* **2011**, *22*, 2117–2131, doi:10.1109/TNN.2011.2172457. [CrossRef] [PubMed]
47. Zhou, G.; Zhao, Q.; Zhang, Y.; Adali, T.; Xie, S.; Cichocki, A. Linked Component Analysis From Matrices to High-Order Tensors: Applications to Biomedical Data. *Proc. IEEE* **2016**, *104*, 310–331, doi:10.1109/JPROC.2015.2474704. [CrossRef]
48. Zhou, G.; Cichocki, A.; Xie, S. Fast Nonnegative Matrix/Tensor Factorization Based on Low-Rank Approximation. *IEEE Trans. Signal Process.* **2012**, *60*, 2928–2940, doi:10.1109/TSP.2012.2190410. [CrossRef]
49. Yang, J.; Guo, Y.; Yang, Z.; Xie, S. Under-Determined Convolutive Blind Source Separation Combining Density-Based Clustering and Sparse Reconstruction in Time-Frequency Domain. *IEEE Trans. Circuits Syst. I Regul. Pap.* **2019**, doi:10.1109/TCSI.2019.2908394. [CrossRef]

Article

An IoT Platform with Monitoring Robot Applying CNN-Based Context-Aware Learning

Moonsun Shin [1], Woojin Paik [1], Byungcheol Kim [2] and Seonmin Hwang [1],*

[1] Department of Software, Konkuk University, Chungju 27478, Korea; msshin@kku.ac.kr (M.S.); wjpaik@kku.ac.kr (W.P.)
[2] Department of Information and Communication, Baekseok University, Cheonan 31065, Korea; bckim@bu.ac.kr
* Correspondence: smhwang@kku.ac.kr; Tel.: +82-43-8403602

Received: 30 March 2019; Accepted: 30 May 2019; Published: 2 June 2019

Abstract: Internet of Things (IoT) technology has been attracted lots of interests over the recent years, due to its applicability across the various domains. In particular, an IoT-based robot with artificial intelligence may be utilized in various fields of surveillance. In this paper, we propose an IoT platform with an intelligent surveillance robot using machine learning in order to overcome the limitations of the existing closed-circuit television (CCTV) which is installed fixed type. The IoT platform with a surveillance robot provides the smart monitoring as a role of active CCTV. The intelligent surveillance robot, which has been built with its own IoT server, and can carry out line tracing and acquire contextual information through the sensors to detect abnormal status in an environment. In addition, photos taken by its camera can be compared with stored images of normal state. If an abnormal status is detected, the manager receives an alarm via a smart phone. For user convenience, the client is provided with an app to control the robot remotely. In the case of image context processing it is useful to apply convolutional neural network (CNN)-based machine learning (ML), which is introduced for the precise detection and recognition of images or patterns, and from which can be expected a high performance of recognition. We designed the CNN model to support contextually-aware services of the IoT platform and to perform experiments for learning accuracy of the designed CNN model using dataset of images acquired from the robot. Experimental results showed that the accuracy of learning is over 0.98, which means that we achieved enhanced learning in image context recognition. The contribution of this paper is not only to implement an IoT platform with active CCTV robot but also to construct a CNN model for image-and-context-aware learning and intelligence enhancement of the proposed IoT platform. The proposed IoT platform, with an intelligent surveillance robot using machine learning, can be used to detect abnormal status in various industrial fields such as factory, smart farms, logistics warehouses, and public places.

Keywords: IoT platform; intelligent monitoring robot; active CCTV; learning model; machine learning; convolutional neural network

---

1. Introduction

The development of Internet of Things (IoT) technology makes it possible to connect smart objects together through the Internet [1]. Advancements in IoT technologies provide enormous potential for high-quality, more convenient, and intelligent service. Various researches on intelligent IoT service systems are attracting attention due to the development of IoT technology. Recent research shows more potential applications of IoT in information intensive industrial sectors. Various needs, such as automatic setting, autonomous control, and optimal operation, are emerging in addition to inter-object connectivity support in the IoT service system. Although it provides connectivity through internet and automation functions by presetting, it is difficult to maintain stable operation and continuous value creation in the application domain. User monitoring and intervention is needed to resolve these problems. The development and popularization of machine learning and deep learning technologies enable a variety of intelligent services and challenges that previously could not be solved. The intelligent IoT service system is defined as a system that acquires data from the environment, recognizes the situation using the acquired data, and interacts with the user environment according to the service rules and the domain knowledge [2]. Therefore, the accuracy of the context-aware learning model-based on the domain knowledge can influence the quality of the intelligent monitoring service. Many researches on intelligent-robot services regarding the various applications have been in progress [3]. Researches on intelligent robots are able to be applied to service and application based on specific domains such as education, entertainment, life, and manufacturing. We tried to combine intelligent robot service and IoT technology in order to create a new context-aware service.

In this paper we propose an IoT platform with an intelligent monitoring robot which monitors the surrounding environment to figure out the situation and inform the administrator when an abnormal situation occurs. Unlike the existing robot system with a separate server, the intelligent monitoring robot in the proposed IoT platform has not only a server built into it, but also many kinds of devices such as the webcam, the radio -frequency (RF), the ultrasonic sensor, temperature sensor, light sensor, and sound sensor. It is designed to provide convenient use of monitor and control at anytime and anywhere using Wi-Fi network.

Especially, the proposed system performs context-aware learning by using a convolutional neural network (CNN)-based machine learning for context-aware learning. CNN is a method of machine learning optimized for image learning because it can input two-dimensional structure. CNN has proven its superior performance in extracting high-level abstracted features from images and recognizing objects in an optimal way.

This paper is organized as follows. Section 2 briefly describes the related works and Section 3 presents the framework of an IoT platform with a monitoring robot. We designed the CNN model for context-aware service as outlined in in Section 4 and analyzed the experimental results as seen in Section 5. Finally, we describe the conclusions in Section 6.

2. Related Works

Many efforts have been conducted toward employing IoT technology in the various industrial field to acquire data, process data timely, and distribute data wirelessly [1]. In recent years, the CCTV video-surveillance system has been introduced into various industrial fields and it has developed into a network-based CCTV or an intelligent-CCTV. The intelligent-CCTV system has been evaluated for its ability to monitor situations very effectively as it can detect the characteristics of an object or a person automatically. Nevertheless, most CCTV systems have been installed in fixed positions, and send images to a central server [4]. Therefore, immediate response to risks and anomalies is difficult because monitoring can only be performed in the control center after the images have been sent to a remote server. It is possible for active CCTV with IoT to perform real-time, context-aware, and immediate response. Recently, the need for context awareness to serve as an intelligent service in ubiquitous environments has increased with the development of a variety of sensor technologies [5]. Many kinds of techniques, such as machine learning, Bayesian network, data mining, and collaborative filtering, are applied for the construction of context-aware models to provide customized intelligence services in a variety of domains [5].

Although various attempts have been made to provide a context-aware service, most of them are being developed as a monitoring framework [3]. In [6], they applied the CNN-scheme-based optical camera communication system for intelligent Internet of vehicles. They used CNN for precise detection and recognition of light-emitting diode patterns at long distances and in bad weather conditions [6]. It can be used to employ CNN in various ways. We propose to employ CNN in the proposed IoT platform with an intelligent surveillance robot and it can be utilized in various fields of surveillance.

A context-aware service can recognize the circumstances, and then provide an appropriate service according to the environment [7]. Recently, the need for a context-aware service as the intelligent service in ubiquitous environments according to the development of sensor technologies has increased [8]. Above all, a context-aware model must be constructed before the context-aware service can be applied. Various techniques such as machine learning, collaborative filtering, and Bayesian network can be used to build context-aware models to provide customized intelligence services in a variety of domains [9]. An ontology-based context model will be able to describe a context semantics method which is independent of system or middleware [9]. However, in the case of a context-aware learning model for image recognition or pattern recognition, there are limitations to an ontology-based context-aware model [10].

In order to overcome these limitations of an ontology-based context-aware model, we employed convolutional neural network (CNN)-based machine learning (ML), which was optimized for image- or pattern-recognition [11,12]. CNNs have developed significantly in recent years and are being used in a variety of areas, such as image and pattern recognition, natural-language processing, video analysis, and speech recognition. The improved network structures of CNNs lead to memory savings and reduced computational complexity and, at the same time, offer better performance for numerous applications. A CNN is composed of a series of layers, wherein each layer describes a specific function. The neuron structure of the artificial neural network is shown in Figure 1. In Figure 1, x is the input signal, W is the weight, b is the bias, f is the activation function, and o is the output. The active function f can be used as a sigmoid function, a hyperbolic tangent function, or a ReLU (rectified linear unit) function. In the case of image classification, the ReLU function has recently been used more than other functions because it shows better performance.

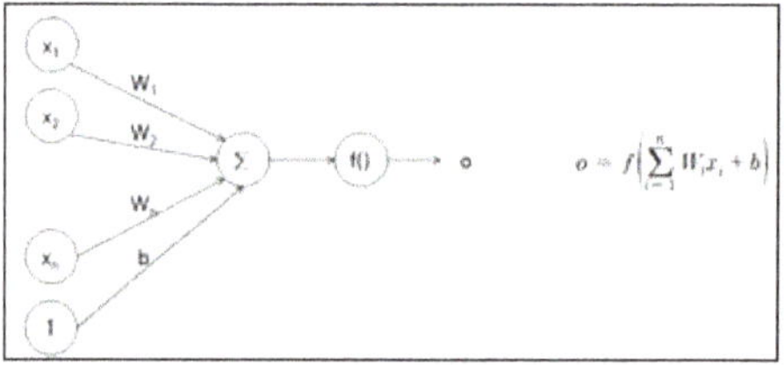

Figure 1. Neuron structure of artificial neuron network (ANN).

To find the optimal weight and bias during learning, the differential is used to get the slope, and the amount of change in the weight is an estimate of the slope. The learning rate is a value that determines how much the parameter value is updated in one learning cycle, and must be set when neural network modeling has been carried out. The learning rate could be a value between 0 and 1. The smaller the learning rate, the slower the speed of learning. If it is set large, the learning speed can be increased, but the neural network may become unstable. Because the neural network model learns the training data excessively in order to decrease the error rate, side effects like overfitting occur and cause it to increase the error rate for actual data. To avoid overfitting, a drop-out technique can be used, which does not use all the nodes of the neural network, but rather selects some of the nodes at random [13]. Whenever weights are updated, the nodes are randomly reconstructed, and learning is performed through them. In the case of using fixed constants as a learning rate, the learning may not be performed properly. To solve these problems Momentum, AdaGrad (adaptive gradient), and Adam (adaptive moments) algorithms have been proposed. Momentum algorithm adds a momentum term to the slope and applies the update more strongly when the slope is in the same direction as the momentum [14]. In our study we adopted ADAM to overcome the overfitting problem. When modeling CNN using a backpropagation algorithm, the performance depends on the initial value of the weight. Preliminary training can be performed using a restricted Boltzmann machine (RBM) or autoencoder to obtain the initial value of the appropriate weight. Hinton proposed the RBM which consists of one input layer and one hidden layer [14]. The initial value of the weight in autoencoder is obtained by preliminarily training performed in each layer of the neural network using the unsupervised learning algorithm. CNN (convolution neural network) is one of the deep learning methods and is used to analyze image data and classify it according to its features [14]. Each layer of CNN has a function o for extracting and learning features by applying a filter to the input image. The CNN is effective in learning on image recognition. The architecture of the CNN is shown in Figure 2.

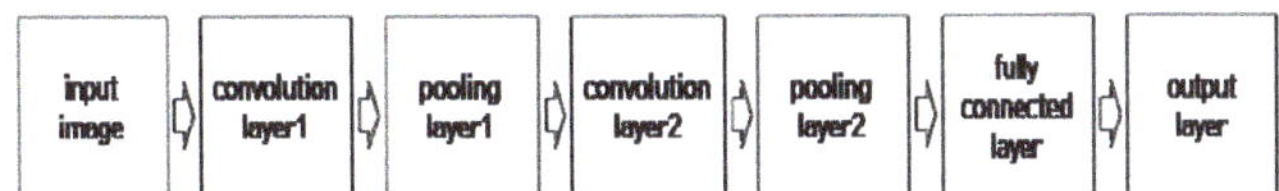

Figure 2. Architecture of convolutional neural network (CNN).

As shown in Figure 2, CNN generates the output from the input image through the convolution layer and the fully connected layer. The convolution layer consists of several convolution layers and pooling layers. The convolution layer generates the convolution output using the input image and the filter and generates the feature map by applying the activation function to the convolution output. The pooling layer generates the output image by reducing the dimension of the feature map using the pooling function. The output image of the last pooling layer is used as the input of the fully connected layer. Yann LeCun developed LeNet5 in 1998 using CNN [15]. This technique has an effect on number recognition. It receives 32 × 32 image and generates output through three convolution layers, two pulling layers, and one fully connected layer. AlexNet, released by Krizhevsky, Sutskever, and Hinton at ILSVRC-2012, was awarded first prize with an error rate of 15.3%, which was remarkably excellent compared to second prize with a 26.2% of error rate. AlexNet consists of five convolution layers, one pooling layer, and three fully connected layers, using two GPUs [16]. ResNet has demonstrated that networks can be deepened to a maximum of 152 layers, and verified better results than fewer layers [17].

These previous works have found that as the depth of the layers becomes deeper, the accuracy of learning is improved in deep learning [18,19]. In our study, we extended LeNet-5 model for context-aware learning of IoT-based intelligent monitoring.

3. Framework of an IoT Platform with an Intelligent Monitoring Robot

In this section, we present a system architecture of an IoT platform with an intelligent monitoring robot by applying a CNN-based context-aware learning model we designed. Figure 3 shows the framework of an IoT platform with an intelligent monitoring robot. As shown in the Figure 3, the IoT-based context-aware system must have its own server and communicate with the web or app client using Wi-Fi. The basic functions required for IoT-based intelligent monitoring systems are learning of situational awareness and real-time recognition process. In addition, in order to provide a notification service for monitoring results and for abnormality detection in real time, it is necessary to make real-time communication always possible by applying IoT technology. The IoT-based context-aware system can be implemented as a fixed or portable type. In the latter case, autonomous navigation using a line tracing or map of a specific area is required.

The proposed system needs to notify the situation through a smart phone and perform image processing functions for the detection of abnormal status. For this the system needs a web server, DB server, and DVR server to be constructed on the main controller in order to search the stored images at the remote site. These servers need to be capable of image storage, event storage, search, and then the web- or app-client will be able to control the movement of the robot for remote users and to receive images. The data transmitted from the attached sensor should be stored in the proposed IoT-based system and be displayed in real time on the remote web browser or app client. The image receiving function and the robot control function need to be supported for the app client, because it is necessary for the user to receive images and to control the movement of the robot by providing the management of a real time context-aware service.

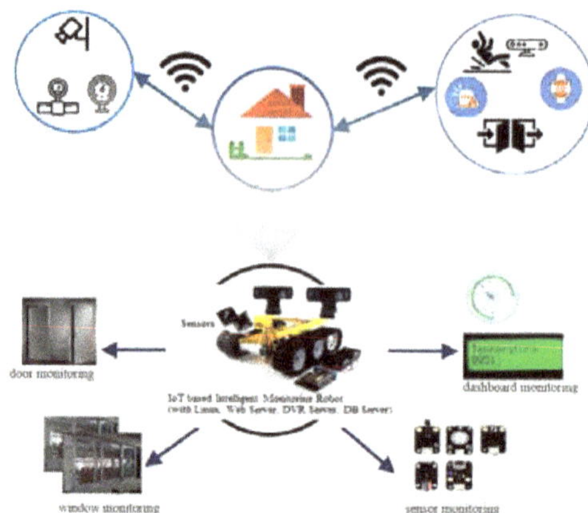

Figure 3. Framework of an IoT-based intelligent monitoring robot.

Figure 4 shows the architecture of IoT platform with an intelligent monitoring system that performs context-aware learning. An IoT platform with a context-awareness system should be able to distinguish between abnormal status and normal status from the monitored sensor value or images from a webcam by performing learning and real-time recognition processes on normal states to check for abnormal states. Also, it must have a variety of sensors that can measure indoor air conditions, such as room temperature and CO, and must be able to perform photo-taking functions to monitor the situation at a specific location. The sensors of the ultrasonic wave, the temperature, the light, the carbon monoxide, and the vibration will be used to detect the situation information of a predetermined area. It also should have a repository that stores the sensing values or images from a webcam while monitoring the environment. We constructed a web server, database server, socket server, and DVR server not only in the IoT-based monitoring robot but also in the backend server for backup and CNN-based context-aware learning.

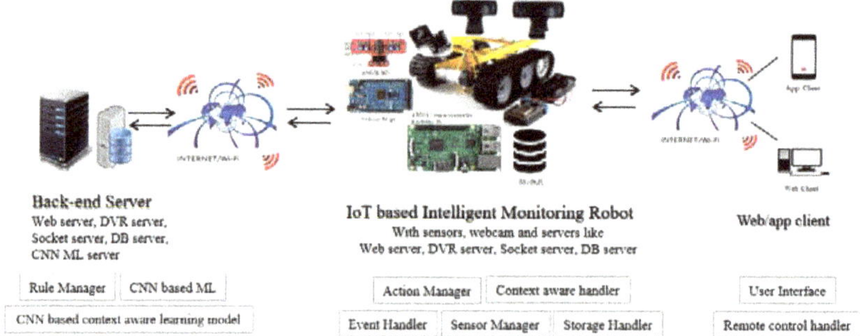

Figure 4. Architecture of the IoT platform with monitoring robot applying CNN machine learning (ML).

Since the server itself is built in the body of the robot, it is able to check the streaming images and sensor value by using the remote client. Images are stored with date time, so it is possible to retrieve the image of the specific date and time by the client. The intelligent monitoring robot of the proposed IoT platform has a motor driver and servomotor, for moving the motor inside, and a infrared sensor for sensors and line tracing are connected to a separate battery and to an Arduino pin. Arduino is capable of serial communication with Raspberry Pie via USB port. Two webcams are used, which are streamed via the Mjpg-streamer. The one in the bottom is only used for streaming, and the other one attached to the servo motor can be used for streaming and image shooting. The hardware device could generate and send various sensor-values and image streaming, and then Node.js web server can parse the received sensor values and send them to the client. After being stored in the database, alarms can be generated abnormality is detected. The web or app client can request the client page via TCP/IP and retrieve the image of the desired date stored in the database. It can request streaming images from a webcam streaming server and receive the contextual information and sensor values in real time via sockets.

The software architecture of the IoT platform with an intelligent monitoring system is shown in Figure 5. It consists of five components: Action Manager, Event Handler, Storage Handler, Reasoning engine, and Rule Engine. Action Manager carried out actions like line tracing for monitoring.

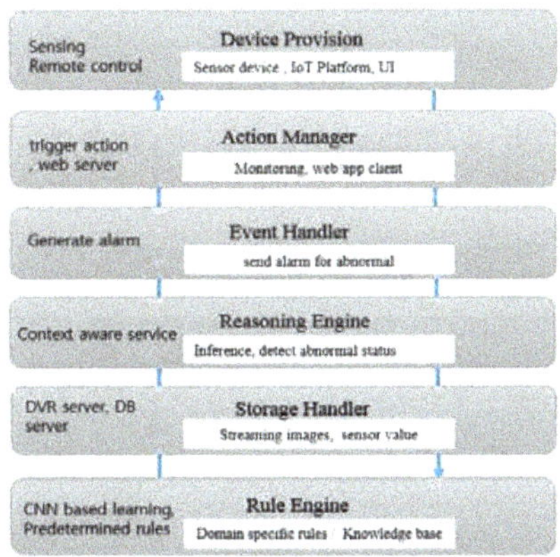

Figure 5. Architecture of the IoT platform with an intelligent monitoring system.

Event Handler generates alarms when abnormal states are detected and Storage Handler stores the images from the webcams and values from sensors. Reasoning engine performs the functions to figure out whether there is an abnormality in situation information such as sensor values and images. Rule Engine performs CNN-based ML for context-aware learning about image context information and builds a knowledge-base with predefined rules. Incremental learning of images that can be gotten continuously by webcam provide guarantees for continuous context learning. Thus, it is possible to make the knowledge-base update incrementally.

Figure 6 shows the sequence diagram of the IoT platform with an intelligent monitoring system. Each situation is perceived according to the sensed values from sensors and a context-aware service is performed accordingly.

If the room temperature is out of the range of the predetermined value, an alarm is generated. Also, if the concentration of carbon monoxide is high, an alarm indicating that the air condition is improper is generated. When providing a surveillance and context awareness service according to images taken by the camera, an alarm is generated when a window or a door is opened.

It also shows the process of providing a continuous monitoring and context-aware service, according to whether the gas valve is locked or not, and whether or not there is a fire extinguisher at a predetermined position. The specification of the abnormal situation had been defined in advance. If the situation is abnormal, the context alarm information can be actively transmitted to the remote site. It can be utilized in user interfaces of various client environments by using standardized transmission technology-based on Wi-Fi network and TCP/IP protocol. We attached six sensors, a temperature sensor, sound sensor, light sensor, vibration sensor, carbon monoxide sensor, and flame detection sensor, to the robot in order to get the context of the specific environment. Figure 7 shows the implemented robot and graphical user interfaces. For the convenience of remote control and notification of abnormal alert app client was implemented as shown in Figure 7c. The 'View & Control' menu, from the main menu, allows the user to control the robot remotely in real time. The 'Search' menu can be used to search the images stored in specific date or time.

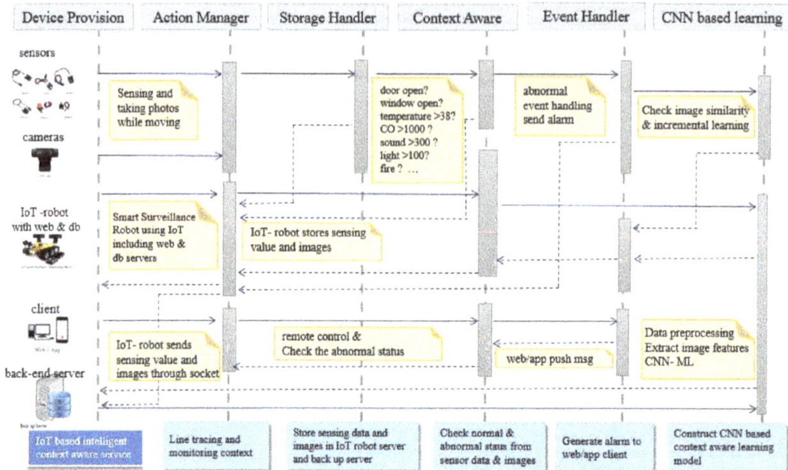

Figure 6. Sequence diagram of the IoT platform with an intelligent monitoring service.

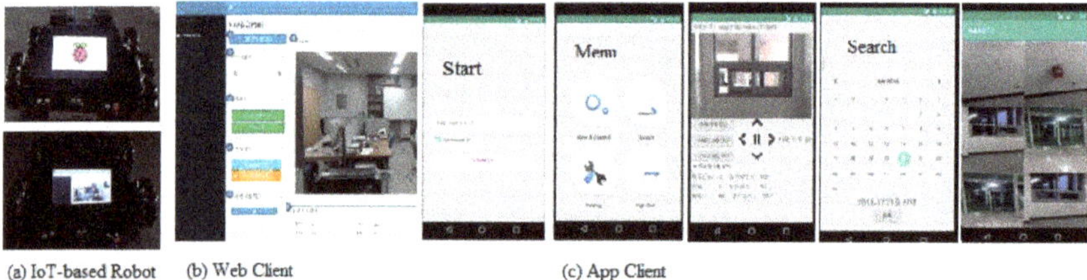

Figure 7. Graphical user interface of IoT platform with an intelligent monitoring robot. (**a**) IoT-based robot; (**b**) Web Client; (**c**) App Client.

When the robot has performed monitoring, photos can be taken by its camera and compared with stored images of a normal state. When an abnormal status is detected, an alarm is sent to the manager via a smart phone. For user convenience, the app-client is able to control the robot remotely. In the case of image-context processing it is useful to apply convolutional neural network (CNN)-based machine learning, which is optimized for image recognition, and can be expected to give a higher performance in the accuracy of learning. A CNN-based context-aware learning model must be constructed in the back-end server. In next section, the design of the CNN-based learning model will be described.

4. CNN-Based Context-Aware Learning Model

In this section, we describe CNN-ML which is adopted for context-aware learning in this paper. First, we have designed the input layer, which defines the type and size of the image input function. The input size varies in accordance to different purposes. For classification tasks, the input size is typically the same size as the training images. However, for detection or recognition tasks, the CNN needs to analyze smaller parts of the image, so the input size must be at least the size of the smallest object in the data set. In this case, the CNN is used to process a (28 × 28)-RGB image. The middle layers, which are the core part of the CNN, consist of convolutional repetitive blocks, ReLU, and pooling layers. The convolutional layers are a set of filter weights which are updated during network training.

The ReLU layer adds non-linear functions to the network. The pooling layers downsample data as they flow through the network. A deeper network can be created by repeating these basic layers.

Weights and activation functions are applied to the convolution output image of each channel. In the pooling layer, an output image with a size of $12 \times 12 \times 16$ is generated by applying pulling with a size of 2×2 and a stride of 2 to an output image to which an activation function is applied. The pooling layer causes effective prevention of overfitting by reducing the size of the synthesized multi-layer output image so as to reduce the number of weights and the amount of computation. The image of the last pooling layer is transformed into a vector to be used as the input of the fully connected neural network. When the filter is applied, the edge information is lost, and the size of the output image is reduced.

In order to compensate for this, the size of the input image and the size of the output image can be made the same by performing a padding process.

We evaluated the constructed CNN model which learns 3000 images taken by robots and classifies them into 10 situations. The scene captured by the robot stopped at a pre-set location inside the building is the state of opening/closing of the outer door, the inner door, and the window. It extracts ten images that distinguish between closed state and open state to be used as training data and test data. Figure 8 shows 10 types of images shot by the robot.

Figure 8. Images of opened/closed of door and window.

For the 640×480 color image taken at each designated location, the robot selects only the center and converts it to 480×480. After extracting the binary image through the grayscale image, it must be resized to 28×28 size and can be used as CNN input. The result of this process is shown as in Figure 9, which presents the original image in (a), the crop image in (b), the gray image in (c), and the binary image in (d).

Figure 9. Convert process of original image. (**a**) Original image; (**b**) crop image; (**c**) gray-scale image; (**d**) binary image.

The architecture of CNN model proposed in this paper is a variant of LeNet-5, and it consists of two convolution layers, two pooling layers, and two fully connected layers. It extracts features from the input image in a convolution layer which is generated by using a convolution of the input image and the weight. Weights are used as filters, and 3×3 filters or 5×5 filters are used for image learning in the proposed CNN model. The filter used in the first convolution layer is $5 \times 5 \times 32$ and receives a

$28 \times 28 \times 1$ input and produces a $28 \times 28 \times 32$ output. Next, as the maximum value pooling step, stride 2 is applied at the maximum value of the 2×2 window to generate the output of $14 \times 14 \times 32$.

In the second convolution layer, a $14 \times 14 \times 32$ input is used to generate a $14 \times 14 \times 64$ output using a $5 \times 5 \times 64$ filter, and as an input, a $7 \times 7 \times 64$ output is generated in a second maximum value pooling. In the first fully connected layer, $7 \times 7 \times 64$ (3136), which is the output of the second pooling layer, is changed to 1024 one-dimensional output. In the second fully connected layer, 1024×10 is generated, and then the final output layer is determined as 10 kinds of status using the softmax function. For example, a sigmoid function, a hyperbolic tangent function, or a ReLU (rectified linear unit) function can be used as an activation function in each convolution layer. Recently, the ReLU function has been used much more because of its high performance. We also use the ReLU activation function in this study. The ReLU function is simply defined as Equation (1).

$$f(x) = max(0, x) \tag{1}$$

The ReLU function is a line with a slope of 1 if $x > 0$ and a slope equals 0 if $x < 0$. The architecture of extended CNN used in our study is shown in Figure 10.

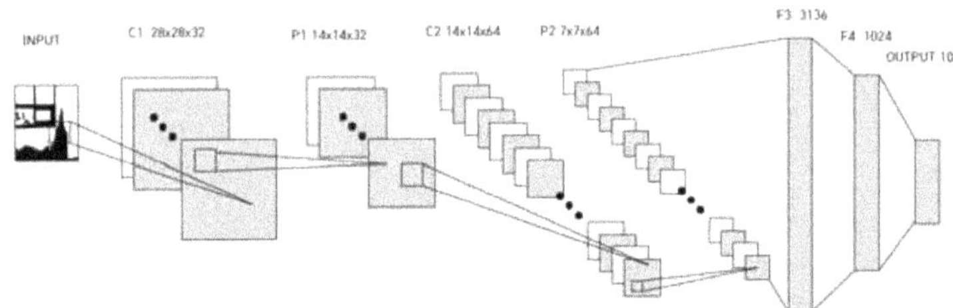

Figure 10. Architecture of extended CNN.

5. Experimental Results and Discussion

The details of the extended CNN model, which consists of 2 convolution layers, 2 maximum pooling layers, and 2 full connection layers, is shown in Table 1. The active function uses ReLU and the input image is converted into a 28×28 size monochrome image.

Table 1. Details of extended CNN model.

Layers	Parameter	Output Size
Input	-	28×28
Conv1	Filter Size: 5×5 Kernel: 32	$28 \times 28 \times 32$
Max Pool1	Filter Size: 5×5 Kernel: 2	$14 \times 14 \times 32$
Conv2	Filter Size: 5×5 Kernel: 64	$14 \times 14 \times 64$
Max Pool2	Filter Size: 5×5 Kernel: 2	$7 \times 7 \times 64$
FC1	Node: 3136	1024
FC2	Node: 1024	10

We used 3000 images taken by the robots for the experiments. Among them 2700 images were used for learning data and 300 images were used for the test data. The learning process was performed through 10,000 epochs totally, and dropout was applied to reduce overfitting while learning.

We adopted the cross-entropy function as a loss function and the ADAM for optimization algorithm. The learning rate is set at 0.05 and converts 1024 features into 10 classes (One-hot Encoding). The CPU used in the experiment is Intel i9-9900K (3.6 GHz) and the GPU is GeForce RTX2080Ti. Figure 11 shows the part of the data used in the learning and as shown in the figure status of the doors and the windows is closed or not.

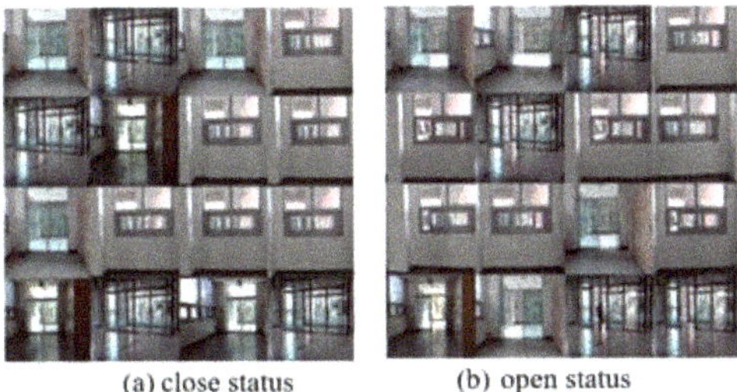

Figure 11. Images of windows and doors used in CNN Learning: (**a**) close status of doors and windows (**b**) open status of doors and windows.

Experimental results showed high recognition rate and it was demonstrated that the change of the dropout was not much influence on accuracy. For the comparison of optimization method, we demonstrated two optimization algorithms AdaGrad and ADAM and verified ADAM has better performance in learning accuracy than AdaGrad.

We tried to check accuracy according to epochs of 1000, 3000, 5000, and 10,000, and were able to verify the accuracy of both algorithms for each epoch. The accuracy rates of each algorithm are shown in Table 2 and Figure 12. In the case of ADAM, the accuracy was 0.9725 for 1000 epochs and 0.9911 for 10,000 epochs. Experimental results showed that 10,000 epochs using Adam optimization showed the highest performance. In experimental results of AdaGrad optimization, the accuracy was 0.8518 for 1000 epochs and 0.9318 for 10,000 epochs. As shown in Table 2, it was figured out that two algorithms had significantly different performance in learning accuracy.

Table 2. Accuracy rate of two algorithm.

Optimization	Epochs	Accuracy
ADAM	1000	0.9725
	3000	0.9836
	5000	0.9882
	10,000	0.9911
AdaGrad	1000	0.8518
	3000	0.8998
	5000	0.9216
	10,000	0.9318

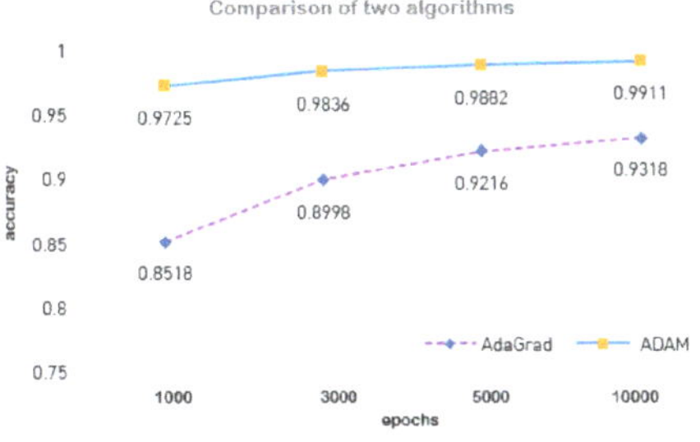

Figure 12. Accuracy comparison of two algorithms: ADAM and AdaGrad.

For the ADAM algorithm, the accuracy graphs were shown in Figure 13 with epochs of 1000, 3000, 5000, and 10,000, respectively. The graphs in Figure 14 showed the accuracy of AdaGrad according to each epoch.

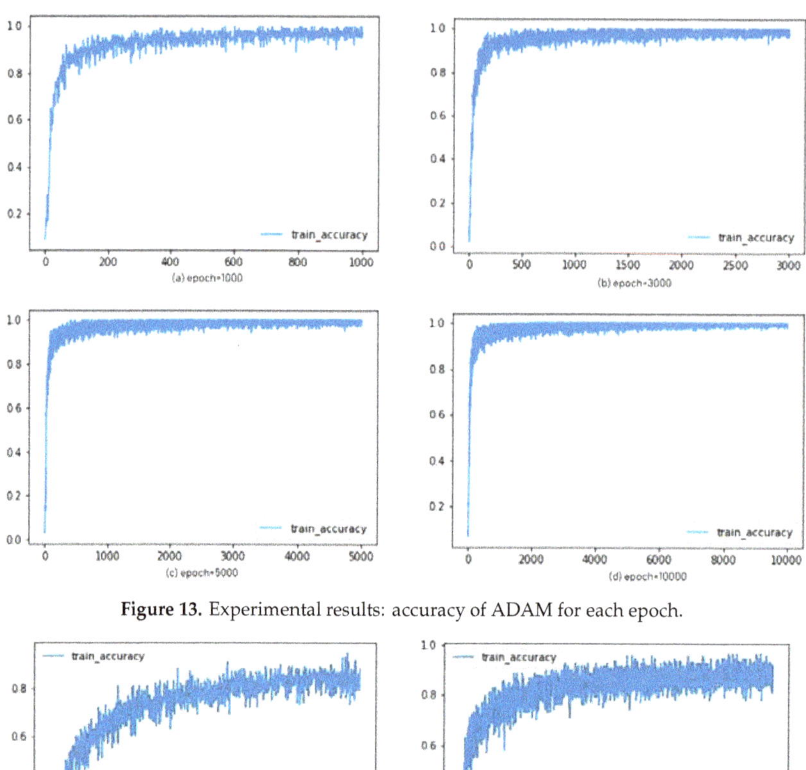

Figure 13. Experimental results: accuracy of ADAM for each epoch.

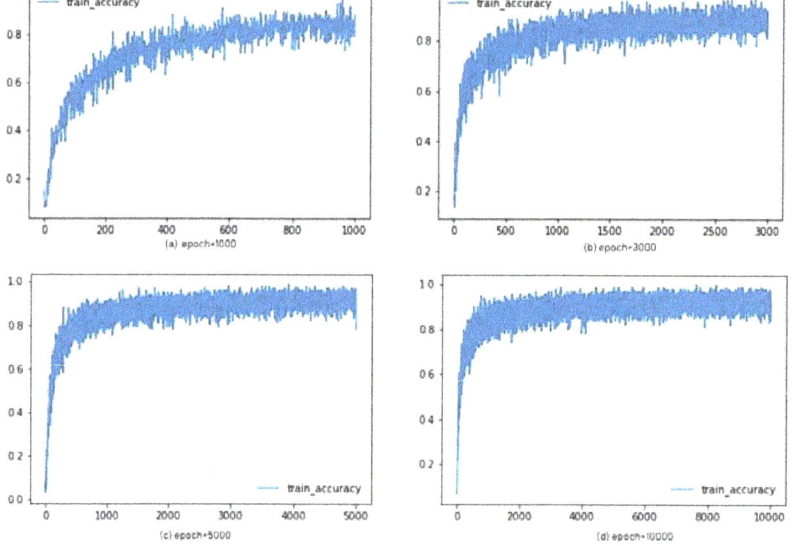

Figure 14. Experimental results: accuracy of AdaGrad for each epoch.

6. Conclusions

IoT technologies bring innovations to wide intelligent robot services in the real-time applications domain. In this paper, we proposed an IoT platform with an intelligent monitoring robot, which can perform functions such as autonomous driving and situational awareness, real-time video transmission, and context-awareness of abnormal situations. We also designed the CNN model for an IoT-based intelligent monitoring robot to support enhancement of context-aware service.

The intelligent monitoring robot in the proposed IoT platform can be used as an active CCTV including an IoT server to overcome the problems of existing fixed CCTV. CNN-based ML is used to figure out whether the monitoring images were of normal or abnormal status in the case of image context. Servers were implemented in the proposed monitoring robot itself, which could perform real-time communication, processing sensing values, and shooting images from a webcam while monitoring. A CNN-based ML server was constructed in the back-end server for context-aware learning. We adopted the cross-entropy function as a loss function and the ADAM for optimization algorithm. The recognition rate was 0.9911 in experimental results.

The contribution of this paper is to improve the accuracy of context-aware learning for an IoT-based active-surveillance robot by applying CNN. The developed IoT-based monitoring robot can be used for rapid resolution of an abnormal situation of an image context in many areas such as the prevention and detection of intrusion, environment pollution, and potential disasters in a variety of fields. We are going to study to improve context-aware learning and to adapt it to actual situations such as in a factory, building, or home environment for practical use. However, the operation time of the robot was only about 5 h, and we exposed that there is a problem with the battery. In future work, we are going to study to improve the performance of the developed robot and to ensure battery efficiency.

Author Contributions: The work described in this article is the collaborative development of all authors. M.S. and S.H. contributed to the conceptualization, formal analysis, experiments, validation, and writing of the manuscript. B.K. performed the data collection and preprocessing, formal analysis, and revision of the manuscript. W.P. contributed to methodology, review, and editing. M.S. contributed to the funding acquisition.

Funding: This paper was funded by Konkuk University in 2018.

Acknowledgments: Seonmin Hwang is corresponding author.

Conflicts of Interest: The authors declare no conflict of interest.

References

1. Boyi, X.; Li Da, X.; Hongming, C.; Cheng, X.; Jingyuan, H.; Fenglin, B. Ubiquitous data accessing method in IoT-based information system for emergency medical services. *IEEE Trans. Ind. Inform.* **2014**, *10*, 1578–1586. [CrossRef]

2. Youngcheol, G.; Joochan, S. Context Modelling for Intelligent Robot Services using Rule and Ontology. In Proceedings of the 7th International Conference on Advanced Communication Technology, ICACT 2005, Phoenix Park, Korea, 21–23 February 2005; Volume 2, pp. 813–816.

3. Moonsun, S.; Myeongcheol, K.; Younjin, J.; Yongwan, J.; Bumju, L. Implementation of context-aware based robot control system for automatic postal logistics. *Stud. Inform. Control* **2013**, *22*, 71–80.

4. Design and Construction of Intelligent Monitoring System for CCTV. *IITTA, ICT Standard Weekly*, 30 September 2013.

5. Daniel, R.; Andries, E. Social network as a coordination techniques for multi-robot systems. In *Intelligent Systems Design and Applications*; Abraham, A., Franke, K., Koppen, M., Eds.; Springer: Berlin/Heidelberg, Germany, 2013; pp. 503–513.

6. Amirul, I.; Md, T.H.; Yeong, M.J. Convolutional neural network scheme–based optical camera communication system for intelligent Internet of vehicles. *Int. J. Distrib. Sens. Netw.* **2018**, *14*, 1–19.

7. Yoongu, K.; Hankil, K.; Sukgyu, L.; Kidong, L. Ubiquitous Home Security Robot Based on Sensor Network. In Proceedings of the 2006 International Conference on Intelligent Agent Technology, Hong Kong, China, 18–22 December 2006; pp. 2060–2065.

8. Claudio, B.; Oliver, B.; Karen, H.; Jadwiga, I.; Daniela, N.; Anand, R.; Daniele, R. A survey of context modelling and reasoning techniques. *Pervasive Mob. Comput.* **2010**, *6*, 161–180.

9. Hyun, K.; Minkyoung, K.; Kangwoo, L.; Youngho, S.; Joonmyun, C.; Youngjo, C. Context-Aware Server Framework for Network-based Intelligent Robot. In Proceedings of the International Joint Conference, SICE 2006, Busan, Korea, 18–21 October 2006; pp. 2084–2089.

10. Moonsun, S.; Byungcheol, K.; Seonmin, H.; Myeongcheol, K. Design and implementation of IoT-based intelligent surveillance robot. *Stud. Inform. Control* **2016**, *25*, 421–432.

11. Xiao-chi, W.; Jie, X.; Zhi-gang, F. Design of smart space context-aware system framework. *Inf. Technol. J.* **2013**, *12*, 5616–5620. [CrossRef]

12. Sangkyun, K.; Yongsoo, J.; Youngmi, L. Sensible media simulation in an automobile application and human responses to sensory effects. *ETRI J.* **2013**, *35*, 1001–1010.

13. Hinton, G.; Osindero, S.; Teh, Y. A fast learning algorithm for deep belief nets. *Neural Comput.* **2006**, *18*, 1527–1554. [CrossRef] [PubMed]

14. Hinton, G.; Salakhutdinov, R. Reducing dimensionality of data with neural networks. *Science* **2006**, *313*, 504–507. [CrossRef] [PubMed]

15. LeCun, Y.; Bottou, L.; Bengio, Y.; Haffner, P. Gradient-based learning applied to document recognition. *Proc. IEEE* **1998**, *86*, 2278–2324. [CrossRef]

16. Krizhevsky, A.; Sutskever, I.; Hinton, G. Imagenet classification with deep convolutional neural networks. *Adv. Neural Inf. Process. Syst.* **2012**, *25*, 1106–1114. [CrossRef]

17. Szegedy, C.; Liu, W.; Jia, Y.; Sermanet, P.; Reed, S.; Anguelov, D.; Erhan, D.; Vanhoucke, V.; Rabinovich, A. Going deeper with convolutions. In Proceedings of the 2015 IEEE Conference on Computer Vision and Pattern Recognition, CVPR, Boston, MA, USA, 7–12 June 2015; pp. 1–9.

18. Simonyan, K.; Zisserman, A. Very deep convolutional networks for large-scale image recognition. In Proceedings of the International Conference on Learning Representations, ICLR 2015, San Diego, CA, USA, 7–9 May 2015; pp. 1–14.

19. He, K.; Zhang, X.; Ren, S.; Sun, J. Deep Residual Learning for Image Recognition. In Proceedings of the IEEE Conference on Computer Vision and Pattern Recognition, Las Vegas, NV, USA, 27–30 June 2016; pp. 770–778.

Article

An Integrative Framework for Online Prognostic and Health Management Using Internet of Things and Convolutional Neural Network

Yuanju Qu [1],*, Xinguo Ming [1], Siqi Qiu [1], Maokuan Zheng [1] and Zengtao Hou [2]

[1]   SJTU Innovation Center of Producer Service Development, Shanghai Research Center for industrial Informatics, Shanghai Key Lab of Advanced Manufacturing Environment, Institute of Intelligent Manufacturing, School of Mechanical Engineering, Shanghai Jiao Tong University, Dongchuan Road 800, Minhang District, Shanghai 200240, China; xgming@sjtu.edu.cn (X.M.); siqiqiu@sjtu.edu.cn (S.Q.); zhengmaokuan@163.com (M.Z.)

[2]   Shenzhen Institutes of Advanced Technology, Chinese Academy of Sciences, 1068 Xueyuan Avenue, Shenzhen University Town, Shenzhen 518055, China; zt.hou@siat.ac.cn

*   Correspondence: iansaqo00@163.com; Tel.: +86-185-6562-0297

Received: 11 April 2019; Accepted: 16 May 2019; Published: 21 May 2019

Abstract: With the development of the internet of things (IoTs), big data, smart sensing technology, and cloud technology, the industry has entered a new stage of revolution. Traditional manufacturing enterprises are transforming into service-oriented manufacturing based on prognostic and health management (PHM). However, there is a lack of a systematic and comprehensive framework of PHM to create more added value. In this paper, the authors proposed an integrative framework to systematically solve the problem from three levels: Strategic level of PHM to create added value, tactical level of PHM to make the implementation route, and operational level of PHM in a detailed application. At the strategic level, the authors provided the innovative business model to create added value through the big data. Moreover, to monitor the equipment status, the health index (HI) based on a condition-based maintenance (CBM) method was proposed. At the tactical level, the authors provided the implementation route in application integration, analysis service, and visual management to satisfy the different stakeholders' functional requirements through a convolutional neural network (CNN). At the operational level, the authors constructed a self-sensing network based on anti-inference and self-organizing Zigbee to capture the real-time data from the equipment group. Finally, the authors verified the feasibility of the framework in a real case from China.

Keywords: prognostic and health management; integrative framework; internet of things; convolutional neural network; conditioned-based maintenance

---

1. Introduction

Prognostic and health management (PHM) [1] is a reliable engineering approach that provides real-time health assessment and predicts its future state by using sensing technologies, machine learning, failure physics, etc. The main goal of PHM technologies is to provide the real-time health state of machines in order to improve the machine's performance by taking proactive actions including diagnostics and prognostics [2,3]. The classification of prognostic models includes physical models, knowledge-based models, data driven models, and hybrid models [4]. The fault detection and failure progression was solved by the Kalman filter state-space predictor, and fuzzy logic classifiers method in an actuator case [5]. Additionally, PHM is usually studied in a laboratory without considering the influence of aging, the effect of people and a working environment, and the subject is usually a single component like gear, bearing and so on, which does not involve multi-sensor information fusion. In study [6], the authors used a 1D convolutional neural network (CNN) in a structural damage detection system. The prognostic algorithms were an effective method to solve fault prognosis in CBM systems for improving prediction accuracy and precision [7]. Meanwhile, Internet of Things (IoTs) is used in tracking, environment monitoring, sensing, and data collection in PHM [8], which is continually being adopted by the industry [9]. The PHM research was considered only from the application layer not involving the management and value layer.

Therefore, the authors proposed an integrative framework including three levels: Strategic level, tactical level, and operational level to study the online PHM of heavy equipment by using IoTs and a two-layer CNN. The main aim was to provide more added value by effectively and efficiently managing the heavy equipment and using the data from massive sensors. In fact, the usage of massive amounts of sensors and IoTs appears only in recent years, and there are no sufficient data especially for heavy equipment whose lifespan is more than twenty years to train weights of the value from different sensors. The authors designed a self-sensing network based on Zigbee [10], which decreased energy consumption by automatically adjusting the transmitting speed of data according to the distance to the failure threshold. Additionally, this paper only focused on the forward transmission of CNN and the weights of sensors were determined by experts.

The rest of this paper is organized as follows: Related research is reviewed in Section 2. The research methodology is illustrated in Section 3. A case study of PHM was carried out in Section 4. The results of implementing the integrative framework are analyzed in Section 5. Conclusions are presented in Section 6.

2. Literature Review

Prognostics and health management are widely used in the product, manufacturing environments, mainly including the monitoring, diagnostics, and prognostics from components, system, network, and related methods. Therefore, this section investigates different aspects of PHM through the previous research efforts.

2.1. Component Layer of Prognostics and Health Management

A lot of researchers paid attention to components problems in PHM, such as the feature selection and management of rolling element bearings [11]. Byington et al. [12] developed a dynamic model of flight actuator to detect faults and predict failure for flight control actuators. Kacprzynski et al. [13] used statistical models to predict degradation rates of turbine compressor. Mba et al. [14] introduced a classification system of health state for a gearbox by integrating stochastic resonance method and hidden Markov modeling (HMM) method. Wang et al. [15] developed a stochastic degradation model to study the capacity degradation of batteries. Li et al. [16] used an ensemble learning method to predict the health degradation of aircraft engines. The studies usually focused on one type of part and there was no application of integrating a lot of sensors.

2.2. System Layer of Prognostics and Health Management

Some researchers studied the PHM of machines from a system aspect and promoted the synthetic application of sensors. Fitouhi and Nourelfath [17] solved a single machine's integrating problem to provide preventive maintenance. Li et al. [18] developed an ensemble degradation model for engineering systems with multiple sensors using the health index synthesis (HIS) approach. Moghaddass and Zuo [19] proposed an integrated framework for a gradual degrading device using multistate stochastic process. Sensor systems [20] were used in the PHM to monitor operational, environmental, performance-related characteristics. Therefore, the number of sensors was small and the studies only referred to the operational level. Researchers have been making inroads into using emerging technologies to improve current practice, but it is not enough.

2.3. Network Layer of Prognostics and Health Management

The network layer of PHM involves equipment to equipment communication, environmental sensing, online monitoring of key parts, and so on. Internet of Things and wireless sensor networks (WSNs) are widely used to get data in the industry [21]. Data aggregation is the strategy of wireless sensor networks [22], which is used in PHM for the communication of information between the equipment. Internet of Things is a network system that connects equipment with sensors, hardware network, and cloud servers [10,23]. Yang et al. [24] proposed a cloud-based prognostics system which was able to provide a low-cost solution for big data collected from a factory. Xia et al. [25] developed a condition monitoring system based on IoTs to resolve potential weaknesses. Korkua et al. [26] proposed a PHM system based on ZigBee to study rotor vibration under different conditions. Li et al. developed a real-time monitoring system for transport machines by using Radio Frequency Identification (RFID) and Global Positioning System (GPS) [27]. These methods only provided partial function applications in PHM and there is still a lack of multi-layer integrating research about PHM.

2.4. Related Method of Prognostics and Health Management

In past research, Bayesian networks [28], time domain analysis [29], gaussian mixture model [30], logistic regression [31], neural network [32], Kalman filter [33], and other algorithms were used in the PHM. Convolutional neural network is excellent in feature extraction of the data and is usually used as artificial intelligent algorithms through back propagation in the prediction of residual useful life (RUL) or fault recognition of parts, which needs a lot of historical data to train weights. Jing et al. [34] developed an innovative CNN for mechanical diagnosis to learn features directly from vibration signals. Jia et al. [35] proposed a local CNN to directly learn the health conditions of machines. Guo et al. [36] predicted bearings' RUL by proposing a recurrent neural network based on a health indicator. Shao et al. [37] developed a deep CNN for rotating machines to provide accurate diagnosis of a certain part by fusing the monitoring data. Jia et al. [38] proposed an intelligent method based on CNN to predict the health condition of gearboxes and bearings. Chen et al. [39] employed three deep CNN models to identify the fault condition of rolling bearings based on the health condition. With the increasing development of IoTs and CNN in the field of processing massive data and things, more research efforts are needed to adapt to the development of the times [40,41]. Although the researchers promoted the development of PHM by using CNN, the results were usually not good because of the lack of data. To better use the CNN under this condition, the weights of the extracted features had to be determined reasonably by experts at the beginning.

2.5. Motivation and Objectives

Therefore, based on the analysis of the literature review, an innovative framework for PHM that integrates CNN technologies and IoTs into current health management practices was proposed to systematically solve management questions of equipment group with little historical data. This paper aims to provide the systematic and comprehensive framework of PHM for the traditional manufacturers to provide continuous service to their customers, by integrating a two-layer CNN and self-sensing network to deal with the complex state of equipment group. This paper reports the first stage of the development, implementation, and evaluation of the framework which demonstrates how a two-layer CNN and Zigbee network can support the integrative framework for PHM innovation from strategy, tactic and operation levels, although not all of the features in the framework were developed in the current research.

3. Methodology

Following the review of literature, a three-level framework based on CNN and IoTs technology was proposed (Figure 1), including the strategic level of PHM, tactical level of PHM, and operational level of PHM. The operational level of PHM which contained products, sensors, network, and database was in charge of the effective and efficient utilization of hardware by resisting interference and reducing energy consumption. The tactical level of PHM which integrated the advanced software and hardware to effectively and efficiently process data, mainly focused on the functional setting, including the visual management, analysis service, and application integrating function. The strategic level of PHM was in charge of defining goals and value types of different stakeholders, including value-added services and business system optimization.

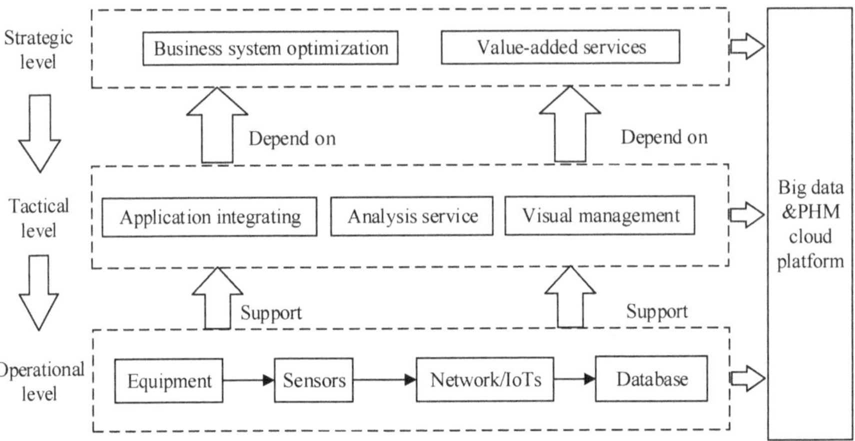

Figure 1. Integrative Prognostics and Health Management framework base on CNN and IoTs.

3.1. Strategic Level Innovation

The strategic level defines the companies' objective of health management and creates a value-added method. Business system optimization and a value-added service helps companies reduce accidents and improve customer satisfaction by predicting service, continuously improving equipment performance in the design phase and extending equipment life.

At this level, HI supported by the tactic level was proposed to reflect the health state. A condition-based maintenance (CBM) method was applied to improve equipment's health state through real-time observation of HI. While HI decreases to a conditioned level, the maintenance will be carried out. The worst part of the equipment and the worst equipment of the batch which were important for the design optimization are also indicated by the framework at this level. A linear function (Equation (1)) was used as the display function by which the range of the HI was magnified from 1 to 100 and the change direction of the HI became positive to the equipment's health state.

$$f(x) = 100 * \left(1 - \frac{x}{T}\right) \tag{1}$$

where, x represents the output value from max pooling layer or convolutional layer and T is the threshold value of x.

3.2. Tactical Level Innovation

At this level, the authors provided application integrating, analysis service, and visual management module for the related stakeholder through CNN with two convolutional layers and two max pooling layers. The CBM-CNN was built to effectively extract the features for the purpose of data analysis, fault correlation analysis, timely detection of potential problems. The expression of the general activation function of CNN is shown in Equation (2).

$$y = f(W \times X + B) \tag{2}$$

where, W represents the weight vector of sensors from equipment; X represents the vector of input value of real-time data from the equipment; B represents the bias of each sensor; f represents the activity function; y represents the health index value of each equipment.

In this paper, the authors use a two-layer CNN to process the data from sensors (Figure 2). The input value X is convoluted through the equipment layer and the output layer and then sent to the display function with the max pooling value from X and equipment layer. The f in the convolution layer (CL) is a rectified linear unit (ReLU) function (Equation (3)).

$$f(x) = \begin{cases} 0, & x < 0 \\ x, & x \geq 0 \end{cases} \tag{3}$$

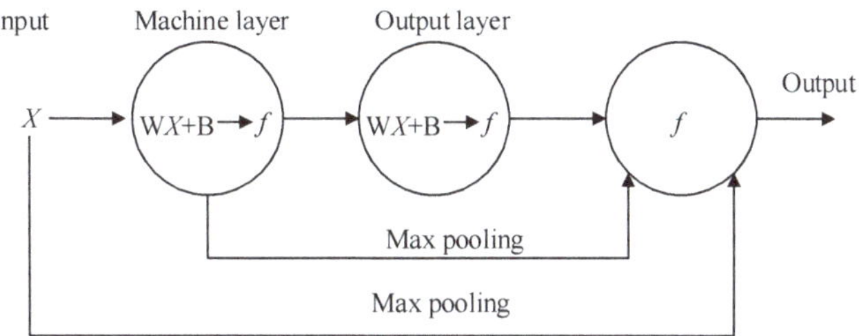

Figure 2. Architecture of the CNN.

The schematic diagram of the CNN is shown in Figure 3. Every row represents an equipment and every column represents the same sensor installed on different equipment. The relevant symbols represent the following meanings:

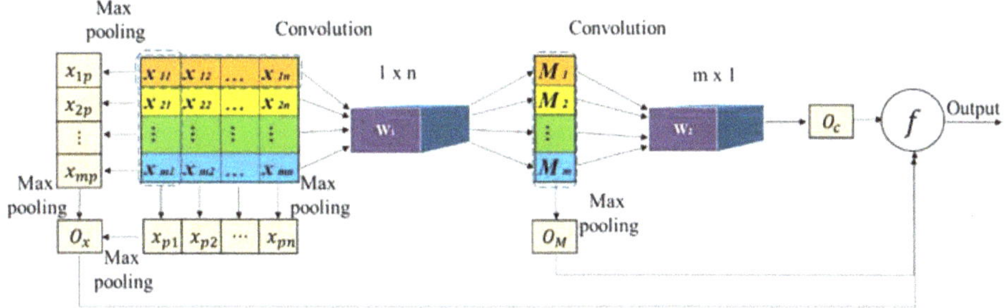

Figure 3. Schematic diagram of the CNN.

x_{mn} is the value from the nth sensor installed on the mth equipment;

M_m represents the synthetic health status of the mth equipment obtained from the first convolution of X and W_1;

x_{mP} is the max pooling value of the mth row, which indicates the worst part of the mth equipment;

x_{Pn} is the max pooling value the nth column, which indicates the worst equipment for the same part;

O_x represents the worst part of all equipment, which is the max pooling value of x_{mP} or x_{Pn};

O_M represents the worst health status of equipment, which is the max pooling value of M_m;

O_c represents the synthetic health status of all equipment, which is the convolutional value of M and W_2;

Through the display function, O_x, O_M, and O_c will be finally displayed in the control center.

3.3. Operational Level Innovation

Technologies of IoTs and smart connection satisfies the requirements of the sensing layer and network layer of PHM. To better resist interference and reduce energy consumption, a self-sensing network based on ZigBee was constructed and the topological structure is illustrated in Figure 4.

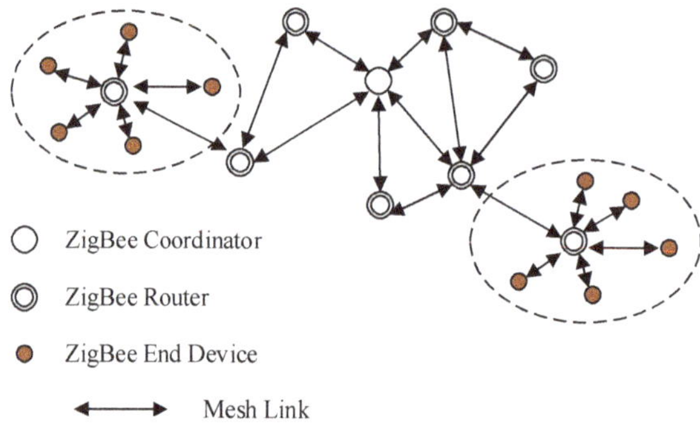

Figure 4. ZigBee topological structure.

In the ZigBee network, there are three kinds of nodes: Coordinator, router and end device. Every network has a coordinator which is the command center. The end devices are installed together with sensors, which are like the dendrites of neurons. The routers are just like the nucleus of neurons which collect data from end devices and communicate with each other. When a certain router breaks down, the corresponding end devices are able to communicate with nearby routers automatically. A self-sensing network is constructed to better balance the data requirements and energy consumption. The transmitting speed varies according to the value x of sensors and frequency f is shown in Equation (4).

$$f = \max(\text{int}(100 * \frac{x}{T})/100, 0.01) * f_0 \tag{4}$$

where f is the transmitting frequency of the node; f_0 is the initial transmitting frequency of the system; T is the failure threshold of the corresponding component; x is the value of the sensor; 'int' means to retain the integer component of the value.

The transmitting speed of this node depends on the real-time value from sensors. When the value is nearer to the failure threshold, the transmitting speed of this node becomes faster so as not to miss important information. The transmitting frequency of the node varies from $0.01 * f_0$ to f_0.

4. Case Study

To validate the feasibility of the proposed approach, a prototype was developed and tested by crane health management (CHM) practitioners. The current version named crane health management systems (CHMS) focuses on cluster health management and only has the ability to provide short-term forecast. The application was built in H company who is a big supplier of port equipment. With the saturation of the market, they wanted to change their role by CHMS innovation so as to create sustainable profits. The framework of CHMS is illustrated in Figure 5.

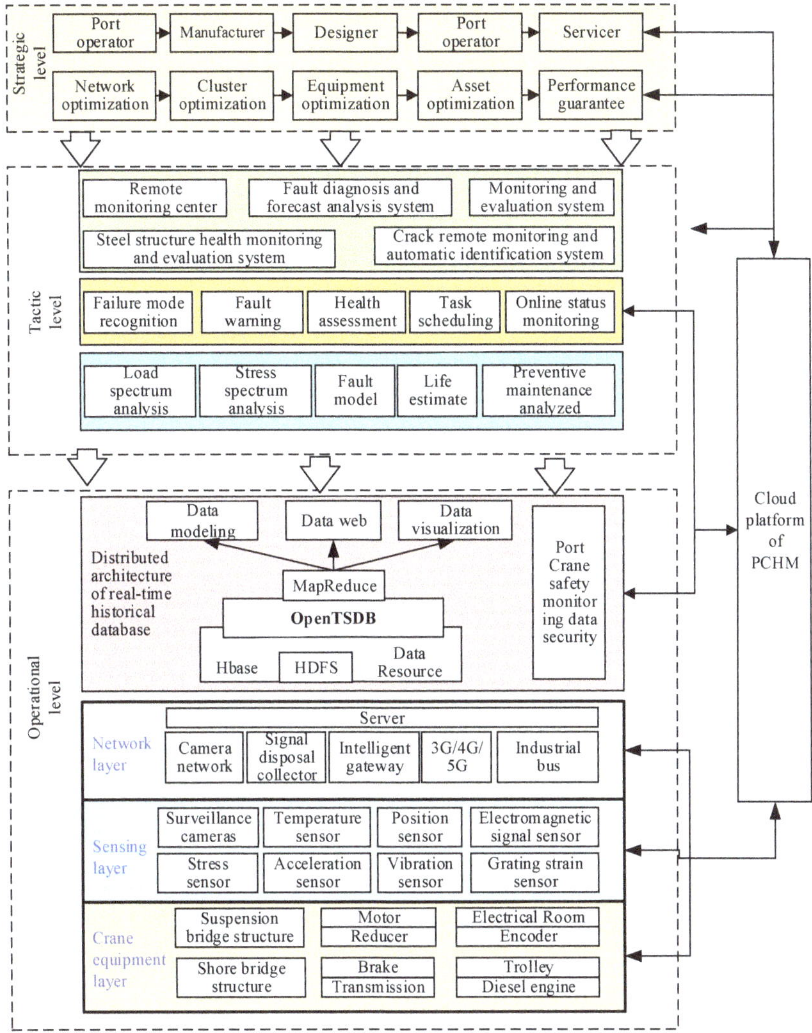

Figure 5. Application of the innovative CHMS framework.

At the operational level, the important parts such as suspension bridge structure, motor and reducer were respectively monitored by corresponding sensors including stress sensor, vibration sensor, acceleration sensor, and so on. In the port, there were 10 cranes and every crane had 67 sensors installed (in Table 1). The designing and choosing principle of the placement of each sensor depended on the historical experience of the fault point and structure characteristic of the equipment. Displacement sensors were installed nearby the holes of the steel structure. Fiber optic acceleration sensors were installed on motors and gear boxes to measure vibration. Full bridge strain gauges were installed on the booms and their connections to measure strain and stress. Fiber Bragg grating temperature sensors were installed in motors and gear boxes to monitor temperature.

Table 1. Sensors installed in each equipment.

Symbol	Description
$X_1 \sim X_{32}$	Displacement sensors need to be arranged at steel structure whose threshold value is 20 mm.
$X_{33} \sim X_{38}$	Stress sensors need to be arranged at different places that are $X_{33} \sim X_{38}$, whose threshold value is 100 MPa
$X_{39} \sim X_{40}$	Strain sensors need to be arranged at different places whose threshold value is 3 mm.
$X_{41} \sim X_{50}$	Axial vibration sensors need to be arranged at different places whose threshold value is 2 mm.
$X_{51} \sim X_{54}$	Radial vibration sensors for main motors need to be arranged at different places whose threshold value is 15 mm.
$X_{55} \sim X_{60}$	Radial vibration sensors for small motors need to be arranged at different places whose threshold value is 10 mm.
$X_{61} \sim X_{67}$	Temperature sensors need to be arranged at gear boxes whose threshold value is 95 °C.

ZigBee wireless network is applied as shown in Figure 6. Each end device was paired with one sensor and all equipment worked with one router. There were still several routers between the cranes and the local servers to amplify the signal and improve the robustness of the network. The data was saved in the local servers and copied to the cloud. The database in the cloud was based on the Hadoop, which was good for dealing with large data and assuring the integrity.

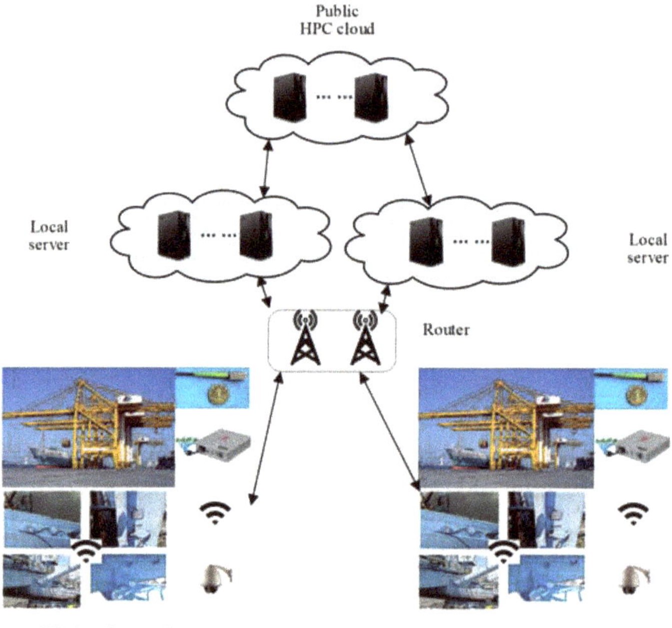

Figure 6. ZigBee wireless network.

At the tactical level, the CNN with two convolutional layers and two max pooling layers were used to analyze the input data including producing HI, finding the worst part and finding the worst equipment. According to the experts, the weights and bias value of CHMS are listed in Table 2.

Table 2. Weights and bias of the CNN for CHMS.

Weights of Sensors	W_1	$w_{11}\sim w_{132}$	$w_{133}\sim w_{138}$	$w_{139}\sim w_{140}$	$w_{141}\sim w_{150}$	$w_{151}\sim w_{154}$	$w_{155}\sim w_{160}$	$\omega_{1161}\sim\omega_{1167}$
		1/20	1/100	1/3	1/2	1/15	1/10	1/30
Bias of sensors	B_1			$b_{11}\sim b_{160}$ 0			$b_{161}\sim b_{167}$ −2.2	
Weights of crane	W_2				$w_{21}\sim w_{210}$ 1/67			
Bias of crane	B_2				$b_{21}\sim b_{210}$ 0			

There was seven new equipment and three old equipment in the port. The input values at a certain time t to the different routers are listed in the Table 3.

Table 3. Matrix of the input values from sensors to different equipment in CHMS.

	X	Ten Equipment in CHMS	
		$X_{M1}\sim X_{M7}$	$X_{M8}\sim X_{M10}$
	$x_1 \sim x_{32}$	1.0	8.0
	$x_{33} \sim x_{38}$	1.7	41.1
	$x_{39} \sim x_{40}$	0.2	1.2
Value of 67 sensors	$x_{41} \sim x_{50}$	0.1	0.7
	$x_{51} \sim x_{54}$	0.7	6.1
	$x_{55} \sim x_{60}$	0.4	3.9
	$x_{61} \sim x_{67}$	48.0	77.0

At the strategic level, business optimization system and value-added service were carried out including performance guarantee, cluster optimization, equipment optimization, and so on (Figure 5). Using CHMS, the managers were able to easily make decisions ahead of time including component replacement, personnel transfer, manufacturing plan, and so on under the CBM principle which was that when the HI decreased by 5, the maintenance would be carried out.

5. Results and Discussion

Replacing the variables by values in Tables 2 and 3, we obtained the values of different layers in CHMS as shown in Table 4.

Table 4. HI of the three layers in CHMS.

Symbol	$M_1\sim M_7$	$M_8\sim M_{10}$
M_m	95.9	60.0
O_x	58.8	
O_M	60.0	
O_C	85.2	

M_m were the machines' HI through displaying function, seven of which were 95.9 and three of which were 60.0. O_x and O_M which represented the HI of the worst part and the worst machine were 58.8 and 60.0 respectively, and O_C which represented the HI of the whole ten machines was 85.2. Because the normal HI of $M_8\sim M_{10}$ was 65, when it declined to 60.0, the principle of CBM was triggered and a maintenance warning was sent to the monitor and mobile terminal ahead of time. The partial interfaces of CHMS are shown in Figure 7, including real-time monitoring interface of port cranes, real-time monitoring interface of steel structures, real-time monitoring interface of a single port crane and real-time monitoring interface of lifting mechanisms. The real-time monitoring interface of port cranes contains Key Performance Indicator (KPIs) of CHMS and overall parameter information of port cranes. The real-time monitoring interface of steel structures displays the value of steel structures in different cranes including stress parameter, displacement parameter and strain parameter. The real-time monitoring interface of a single port crane contains all parameter information that can demonstrate the HI of a port crane. The real-time monitoring interface of lifting mechanisms shows the value of lifting mechanisms in different cranes including vibration parameter, stress parameter, temperature parameter and strain parameter.

(**a**) Real-time monitoring interface of port crane

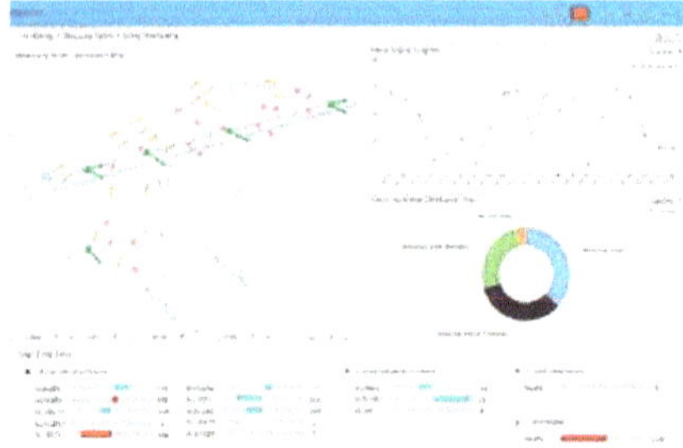

(**b**) Real-time monitoring function interface of steel structure

(**c**) Real-time monitoring interface of single port crane

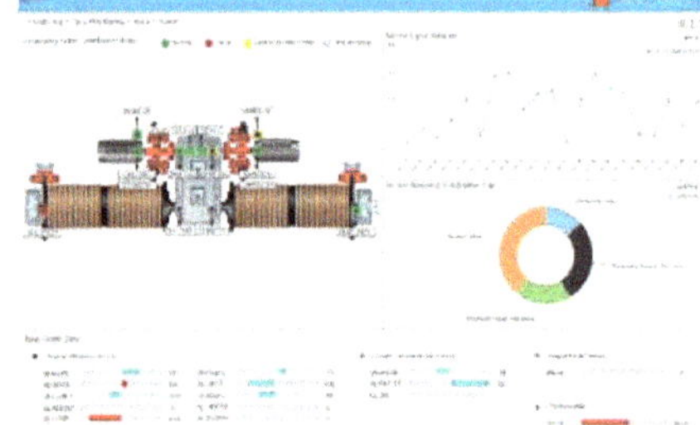

(**d**) Real-time monitoring interface of lifting mechanism

Figure 7. Application interfaces of CHMS.

After application of this framework in CHSM, the three old equipment were dumped and finally the residual useful life of the equipment was prolonged from 5 years judged by experiences to 6 years. Figure 8 shows the variation of health degradation with the application of this framework. Point A is the turning point of health degradation in CHMS. The past curve is based on history data and the expected curve is the prognostic curve based on the traditional method. In the new current curve, CHMS slowed down the health degradation and thus caused the extension of the RUL from t_1 to t_2, which was about 1 year.

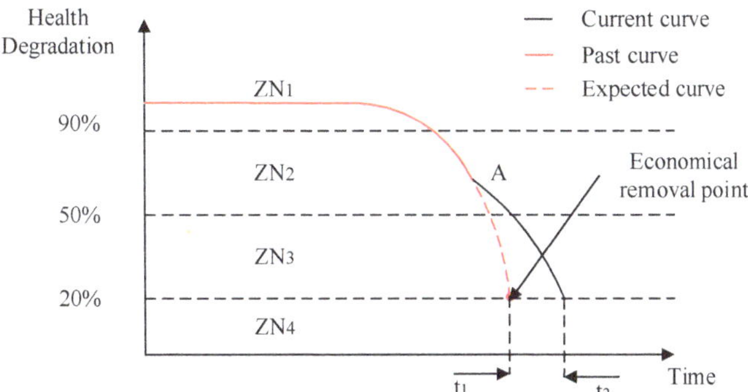

Figure 8. The variation of health degradation. Notes: ZN1 is the safe area; ZN2 is the wearing area; ZN3 is the high-risk area; ZN4 is the scrap area.

Apart from the extension of the RUL, there were other improvements for the stakeholders' value. Table 5 presents the comparison before and after implementing the framework.

Table 5. Comparison before and after implementing the framework in CHMS.

Stakeholder	Item	Before	After	Main Contribution
Company H	Business model	Sell equipment to the port and time-based maintenance	Condition based maintenance and provide health management service, fault prognostic and diagnostic service, breakdowns avoiding service for charges	Business model transition from manufacturer to service provider; Work enthusiasm is incented by reducing personnel and increasing salary; Supply better product based on the history data. Reducing the waiting time.
	Personnel	2	1	
	Salary	Every worker earns RMB 80,000 per year	Every worker gets as much as RMB 90,000	
	Added value	0	Nearly RMB 100,000	
End user	Personnel	3	2	The labor and downtime costs decrease a lot and the productivity rises greatly
	Breakdowns	More than 10 times per year	0	
	Fault rate	About 5% every year	Less than 1%	
	Residual useful life	According to experience, the old equipment's RUL is about 5 years	By the framework, the RUL of the old equipment is prolonged to about 6 years	

With the application of CHMS, more added value is achieved for the manufacturer and the end user. Although the prognostic ability of the system is weak now, with the accumulation of data in such a framework, the training ability of CNN will become true and thus the system can be better in the future.

6. Conclusions

Prognostic and health management is an important method for manufacturers in order to monitor failure precursor, improve product performance, and create added value. Examining the existing literature related to PHM, this paper proposed an integrative framework of PHM based on IoTs and CNN through practical investigation. The framework provided the systematic guidance of PHM for manufacturers from three levels: Strategic level, tactical level and operational level, which would help more companies build a win-win relationship and create more added value, such as the transition from selling products to selling services, continuously improving product performance, and continuously improving customers' satisfaction. At the same time, this paper also plays an exemplary role in showcasing the usage of CNN and IoTs in the fusion of massive sensors. The contributions of this paper were concluded as follows:

- This paper provided an integrative framework for PHM to give a new business model for traditional manufacturers in expanding innovative business model and achieving added value from three levels.

- At the strategic level, the HI was proposed and used with CBM to provide value-added service and business chance referring to optimize product performance and reduce the operation and maintenance cost.

- At the tactical level, the authors developed a two-layer CNN with reasonable weights achieving more added value by effectively and efficiently managing the heavy equipment and using the data from massive sensors.

- At the operational level, this paper proposed the self-sensing network based on Zigbee to realize the monitoring of the real-time data from the equipment group.

- The case study of CHMS was proposed in this paper to check the feasibility of the framework from the value added and prediction of residual useful life of heavy equipment.

Although the proposed integrative framework demonstrates potentials in PHM, there is still more research that needs to be done in the future. To cope with real world practices, more data needs to be collected under this framework to optimize the weights and reveal the relationship between the HI and the RUL through deep learning. Future research will focus on this area.

Author Contributions: Conceptualization, Y.Q. and M.Z.; Formal analysis, Y.Q. and S.Q.; Investigation, X.M. and M.Z.; Methodology, Y.Q. and Z.H.; Project administration, X.M.

Funding: This work was supported by the National Natural Science Foundation of China (Grant No. 71632008) and Major Project for Aero engines and Gas turbines (Grant No. 2017-I-0007-0008, Grant No.2017-I-0011) for the funding support to this research.

Acknowledgments: The author would like to thank SJTU Innovation Center of Producer Service Development, Shanghai Research Center for industrial Informatics, Shanghai Key Lab of Advanced Manufacturing Environment, National Natural Science Foundation of China (Grant No. 71632008) and Major Project for Aero engines and Gas turbines (Grant No. 2017-I-0007-0008, Grant No.2017-I-0011) for the funding support to do this research.

Conflicts of Interest: The authors declare no conflict of interest.

Acronyms

CNN	Convolutional Neural Network
CBM	Condition-Based Maintenance
GPS	Global Positioning System
HI	Health Index
HIS	Health Index Synthesis
HMM	hidden Markov modelling
IoTs	Internet of Things
PHM	Prognostic and Health Management
RFID	Radio Frequency Identification
RUL	Residual Useful Life
ReLU	Rectified Linear Unit
~	means "to"
HPC	High Performance Computing

References

1. Kim, N.-H.; An, D.; Choi, J.-H. *Prognostics and Health Management of Engineering Systems*; Springer International Publishing: Cham, Switzerland, 2017.

2. Teixeira, E.L.S.; Tjahjono, B.; Alfaro, S.C.A. A novel framework to link Prognostics and Health Management and Product–Service Systems using online simulation. *Comput. Ind.* **2012**, *63*, 669–679. [CrossRef]

3. Lee, J.; Wu, F.; Zhao, W.; Ghaffari, M.; Liao, L.; Siegel, D. Prognostics and health management design for rotary machinery systems—Reviews, methodology and applications. *Mech. Syst. Signal Process.* **2014**, *42*, 314–334. [CrossRef]

4. Elghazel, W.; Bahi, J.; Guyeux, C.; Hakem, M.; Medjaher, K.; Zerhouni, N. Dependability of wireless sensor networks for industrial prognostics and health management. *Comput. Ind.* **2015**, *68*, 1–15. [CrossRef]

5. Byington, C.S.; Watson, M.; Edwards, D. Data-driven neural network methodology to remaining life predictions for aircraft actuator components. In Proceedings of the 2004 IEEE Aerospace Conference Proceedings (IEEE Cat. No. 04TH8720), Big Sky, MT, USA, 6–13 March 2004; pp. 3581–3589.

6. Abdeljaber, O.; Avci, O.; Kiranyaz, S.; Gabbouj, M.; Inman, D.J. Real-time vibration-based structural damage detection using one-dimensional convolutional neural networks. *J. Sound Vib.* **2017**, *388*, 154–170. [CrossRef]

7. Khawaja, T.; Vachtsevanos, G.; Wu, B. Reasoning about uncertainty in prognosis: A confidence prediction neural network approach. In Proceedings of the NAFIPS 2005–2005 Annual Meeting of the North American Fuzzy Information Processing Society, Detroit, MI, USA, 26–28 June 2005; pp. 7–12.

8. Atzori, L.; Iera, A.; Morabito, G. The Internet of Things: A survey. *Comput. Netw.* **2010**, *54*, 2787–2805. [CrossRef]

9. Mahesh, S.; Landry, B.; Sridhar, T.; Walsh, K.R. A decision table for the cloud computing decision in small business. *Inf. Resour. Manag. J.* **2011**, *24*, 9–25. [CrossRef]

10. Khan, R.; Khan, S.U.; Zaheer, R.; Khan, S. Future internet: The internet of things architecture, possible applications and key challenges. In Proceedings of the 2012 10th International Conference on Frontiers of Information Technology, Islamabad, India, 17–19 December 2012; pp. 257–260.

11. Liu, Z.; Zuo, M.J.; Qin, Y. Remaining useful life prediction of rolling element bearings based on health state assessment. *Proc. Inst. Mech. Eng. Part C J. Mech. Eng. Sci.* **2015**, *230*, 314–330. [CrossRef]

12. Byington, C.S.; Watson, M.; Edwards, D.; Stoelting, P. A model-based approach to prognostics and health management for flight control actuators. In Proceedings of the 2004 IEEE Aerospace Conference Proceedings (IEEE Cat. No. 04TH8720), Big Sky, MT, USA, 6–13 March 2004; pp. 3551–3562.

13. Kacprzynski, G.J.; Gumina, M.; Roemer, M.J.; Caguiat, D.E.; Galie, T.R.; McGroarty, J.J. A prognostic modeling approach for predicting recurring maintenance for shipboard propulsion systems. In Proceedings of the ASME Turbo Expo 2001: Power for Land, Sea, and Air, New Orleans, LA, USA, 4–7 June 2001; p. V001T002A003.

14. Mba, C.U.; Makis, V.; Marchesiello, S.; Fasana, A.; Garibaldi, L. Condition monitoring and state classification of gearboxes using stochastic resonance and hidden Markov models. *Measurement* **2018**, *126*, 76–95. [CrossRef]

15. Wang, W.; Liu, X.; Cai, F.; Wang, J. Stochastic dynamic modeling of lithium battery via expectation maximization algorithm. *Neurocomputing* **2016**, *175*, 421–426. [CrossRef]

16. Li, Z.; Wu, D.; Hu, C.; Terpenny, J. An Ensemble Learning-based Prognostic Approach with Degradation-Dependent Weights for Remaining Useful Life Prediction. *Reliab. Eng. Syst. Saf.* **2017**, *184*, 110–122. [CrossRef]

17. Fitouhi, M.C.; Nourelfath, M. Integrating noncyclical preventive maintenance scheduling and production planning for a single machine. *Int. J. Prod. Econ.* **2012**, *136*, 344–351. [CrossRef]

18. Li, Y.; Shi, J.; Gong, W.; Zhang, M.; Li, Y.; Shi, J.; Gong, W.; Zhang, M.; Li, Y.; Shi, J. An ensemble model for engineered systems prognostics combining health index synthesis approach and particle filtering. *Qual. Reliab. Eng. Int.* **2017**, *33*, 2711–2725.

19. Moghaddass, R.; Zuo, M.J. An integrated framework for online diagnostic and prognostic health monitoring using a multistate deterioration process. *Reliab. Eng. Syst. Saf.* **2014**, *124*, 92–104. [CrossRef]

20. Cheng, S.; Azarian, M.H.; Pecht, M.G. Sensor systems for prognostics and health management. *Sensors (Basel)* **2010**, *10*, 5774–5797. [CrossRef]

21. Li, W.; Kara, S. Methodology for monitoring manufacturing environment by using wireless sensor networks (WSN) and the internet of things (IoT). *Procedia CIRP* **2017**, *61*, 323–328. [CrossRef]

22. Farhat, A.; Guyeux, C.; Makhoul, A.; Jaber, A.; Tawil, R.; Hijazi, A. Impacts of wireless sensor networks strategies and topologies on prognostics and health management. *J. Intell. Manuf.* **2017**. [CrossRef]

23. Meraghni, S.; Terrissa, L.S.; Zerhouni, N.; Varnier, C.; Ayad, S. A Post-Prognostics Decision framework for cell site using Cloud Computing and Internet of Things. In Proceedings of the 2016 2nd International Conference on Cloud Computing Technologies and Applications (CloudTech), Marrakech, Morocco, 24–26 May 2016; pp. 310–315.

24. Yang, S.; Bagheri, B.; Kao, H.-A.; Lee, J. A unified framework and platform for designing of cloud-based machine health monitoring and manufacturing systems. *J. Manuf. Sci. Eng.* **2015**, *137*, 040914. [CrossRef]

25. Xia, M.; Li, T.; Zhang, Y.; de Silva, C.W. Closed-loop design evolution of engineering system using condition monitoring through internet of things and cloud computing. *Comput. Netw.* **2016**, *101*, 5–18. [CrossRef]

26. Korkua, S.; Jain, H.; Lee, W.-J.; Kwan, C. Wireless health monitoring system for vibration detection of induction motors. In Proceedings of the Industrial and Commercial Power Systems Technical Conference (I&CPS), Tallahassee, FL, USA, 9–13 May 2010; pp. 1–6.

27. Li, H.; Chan, G.; Skitmore, M. Integrating real time positioning systems to improve blind lifting and loading crane operations. *Constr. Manag. Econ.* **2013**, *31*, 596–605. [CrossRef]

28. Ferreiro, S.; Arnaiz, A.; Sierra, B.; Irigoien, I. Application of Bayesian networks in prognostics for a new Integrated Vehicle Health Management concept. *Expert Syst. Appl.* **2012**, *39*, 6402–6418. [CrossRef]

29. Tahir, M.M.; Khan, A.Q.; Iqbal, N.; Hussain, A.; Badshah, S. Enhancing fault classification accuracy of ball bearing using central tendency based time domain features. *IEEE Access* **2017**, *5*, 72–83. [CrossRef]

30. Das, S.; Hall, R.; Patel, A.; McNamara, S.; Todd, J. An open architecture for enabling CBM/PHM capabilities in ground vehicles. In Proceedings of the 2012 IEEE Conference on Prognostics and Health Management, Denver, CO, USA, 18–21 June 2012; pp. 1–8.

31. Mendes, A.C.; Fard, N. Binary logistic regression and PHM analysis for reliability data. *Int. J. Reliab. Qual. Saf. Eng.* **2014**, *21*, 1450023. [CrossRef]

32. Babu, G.S.; Zhao, P.; Li, X.-L. Deep convolutional neural network based regression approach for estimation of remaining useful life. In *International Conference on Database Systems for Advanced Applications*; Springer: Cham, Switzerland, 2016; pp. 214–228.

33. Lim, C.K.R.; Mba, D. Switching Kalman filter for failure prognostic. *Mech. Syst. Signal Process.* **2015**, *52*, 426–435. [CrossRef]

34. Jing, L.; Zhao, M.; Li, P.; Xu, X. A convolutional neural network based feature learning and fault diagnosis method for the condition monitoring of gearbox. *Measurement* **2017**, *111*, 1–10. [CrossRef]

35. Jia, F.; Lei, Y.; Guo, L.; Lin, J.; Xing, S. A neural network constructed by deep learning technique and its application to intelligent fault diagnosis of machines. *Neurocomputing* **2017**, *272*, 619–628. [CrossRef]

36. Guo, L.; Li, N.; Jia, F.; Lei, Y.; Lin, J. A recurrent neural network based health indicator for remaining useful life prediction of bearings. *Neurocomputing* **2017**, *240*, 98–109. [CrossRef]

37. Shao, H.; Jiang, H.; Wang, F.; Zhao, H. An enhancement deep feature fusion method for rotating machinery fault diagnosis. *Knowl. Based Syst.* **2017**, *119*, 200–220. [CrossRef]

38. Jia, F.; Lei, Y.; Lin, J.; Zhou, X.; Lu, N. Deep neural networks: A promising tool for fault characteristic mining and intelligent diagnosis of rotating machinery with massive data. *Mech. Syst. Signal Process.* **2016**, *72–73*, 303–315. [CrossRef]

39. Chen, Z.; Deng, S.; Chen, X.; Li, C.; Sanchez, R.V.; Qin, H. Deep neural networks-based rolling bearing fault diagnosis. *Microelectron. Reliab.* **2017**, *75*, 327–333. [CrossRef]

40. Julka, N.; Thirunavukkarasu, A.; Lendermann, P.; Gan, B.P.; Schirrmann, A.; Fromm, H.; Wong, E. Making use of prognostics health management information for aerospace spare components logistics network optimisation. *Comput. Ind.* **2011**, *62*, 613–622. [CrossRef]

41. Chen, Z.; Yang, Y.; Hu, Z.; Zeng, Q. Fault prognosis of complex mechanical systems based on multi-sensor mixtured hidden semi-Markov models. *Proc. Inst. Mech. Eng. Part C J. Mech. Eng. Sci.* **2012**, *227*, 1853–1863. [CrossRef]

Article

Landslide Susceptibility Assessment Using Integrated Deep Learning Algorithm along the China-Nepal Highway

Liming Xiao [1], Yonghong Zhang [1] and Gongzhuang Peng [2,*]

[1] Department of Information and Communication, Nanjing University of Information Science and Technology, Nanjing 210044, China; 20161118086@nuist.edu.cn (L.X.); zyh@nuist.edu.cn (Y.Z.)
[2] Engineering Research Institute, University of Science and Technology Beijing, Beijing 100083, China
* Correspondence: gzpeng@ustb.edu.cn; Tel.: +86-134-2601-6643

Received: 15 October 2018; Accepted: 12 December 2018; Published: 14 December 2018

Abstract: The China-Nepal Highway is a vital land route in the Kush-Himalayan region. The occurrence of mountain hazards in this area is a matter of serious concern. Thus, it is of great importance to perform hazard assessments in a more accurate and real-time way. Based on temporal and spatial sensor data, this study tries to use data-driven algorithms to predict landslide susceptibility. Ten landslide instability factors were prepared, including elevation, slope angle, slope aspect, plan curvature, vegetation index, built-up index, stream power, lithology, precipitation intensity, and cumulative precipitation index. Four machine learning algorithms, namely decision tree (DT), support vector machines (SVM), Back Propagation neural network (BPNN), and Long Short Term Memory (LSTM) are implemented, and their final prediction accuracies are compared. The experimental results showed that the prediction accuracies of BPNN, SVM, DT, and LSTM in the test areas are 62.0%, 72.9%, 60.4%, and 81.2%, respectively. LSTM outperformed the other three models due to its capability to learn time series with long temporal dependencies. It indicates that the dynamic change course of geological and geographic parameters is an important indicator in reflecting landslide susceptibility.

Keywords: landslide susceptibility; China-Nepal Highway; machine learning; LSTM; remote sensing images

---

1. Introduction

The China-Nepal Highway is a vital land route connecting China and Nepal, which is also an important part of the "One Belt and One Road" development strategy. It is located in the Hindu Kush-Himalayan region—one of the most tectonically active regions of the world. Due to the fragile ecological environment and highly-varying hydrothermal conditions, mountain hazards such as landslides and mudslides take place frequently and have caused severe damage to infrastructure. Thus, it is of great importance to perform the mountain hazard assessment in a more accurate and real-time way. Taking landslide related hazards as the research object, a prediction model is established to assess the susceptibility in this paper.

In the past, disaster information extraction and prediction were mainly based on artificial visual interpretation. Apart from being time-consuming and strenuous, the traditional method also has a limitation in that the measurement process lacks of accuracy and depends heavily on experts' experience. With the development of the computer vision and pattern recognition technologies, it is possible to make the hazard assessment automatic. Synthetic aperture radar (SAR) images have been employed to monitor the surface movement of landslides [1]. Vahidnia et al. [2] applied geographic information systems (GIS) to produce a landslide susceptibility map in which the slope failures that are most likely to happen are displayed. Owing to its high spatial resolution and stereo capability, high-resolution remote sensing images have played an important role in improving the efficiency and accuracy of hazard monitoring [3,4]. The other type of monitoring method is to embed different kinds of sensors related to slope, rainfall, water table level, and other factors into the landslide and sense the dynamic change of signals. Wireless sensor networks are therefore being used to achieve large-scale data collection and transmission [5].

By employing different sensing and monitoring techniques [6–9], multidimensional and multiscale temporal and spatial data can be collected. Based on the data, a variety kind of models and algorithms have been employed in landslide susceptibility assessment. Statistical regression models are typical methods to directly describe the spatial relationships between landslide occurrence and effecting factors [10–12]. Nandi et al. evaluated the multivariate statistical relationship between landslides and various instability factors including slope angle, proximity to stream, soil erodibility, and soil type based on the logistic regression approach [10]. Due to the non-linear condition of hazard prediction, conventional regressive models fail to accurately characterize the causality among variables correctly. Data-driven approaches rely mainly on historical data and do not assume any form of mechanism information, and they have already received much attention in hazard susceptibility assessments, such as support vector machine (SVM), decision tree (DT), neural networks (NN) and so on [13–22]. Liu et al. developed a hybrid BP neural network to assess the geological hazard risk which adopted genetic algorithm (GA) and particle swarm optimization (PSO) to optimize the network connection weights and thresholds [13]. Marjanović modeled the landslide susceptibility assessment problem as a classification problem, and applied SVM to evaluate which category the region belongs to—stable ground, or dormant and active landslides [4]. As expert experience is helpful to improve prediction accuracy, adaptive neuro-fuzzy inference (ANFI) and Bayesian inference are also widely used in susceptibility assessments [23–27]. Vahidnia employs a fuzzy inference system (FIS) to model expert knowledge, and an artificial neural network (ANN) to assess landslide susceptibility by identifying non-linear behavior and generalizing historical data to the entire region [2]. Chalkias used an expert-based fuzzy weighting (EFW) approach to determine the susceptibility level of different regions by weighted linear combination, in which precipitation, slope, and lithology were considered to be the most important conditioning factors [27].

The formation and occurrence of landslides is a complicated evolution process, which is caused by the interaction of multiple instability factors. However, most of the methods consider only the current value of the instability factors while ignoring the factors' evolution feature over time. The recurrent neural network (RNN) can use internal memory units to process arbitrary sequences of inputs, thus making RNNs capable of learning temporal sequence. As a special RNN architecture, LSTM inherits RNNs' good features of sequence learning, and is able to learn the time series with long temporal dependency and automatically determine the optimal result by applying the gate control mechanism. Thus, LSTM has recently attracted wide attention in time series predictions, natural language generation, and so on 28-30]. Ma et al. present a novel LSTM NN to predict travel speed with long time dependencies using microwave detector data. The numerical experiments demonstrate that the LSTM NN outperforms Elman NN, TDNN, and NARX NN in terms of accuracy and stability [28]. Yu developed a transient stability assessment system based on the LSTM network, aiming at balancing the trade-off between assessment accuracy and response time [29]. To our knowledge, Mezaal was the first to use RNN in automatic landslide detection from high-resolution airborne laser scanning data, with an accuracy of more than 80% [30]. In this paper, LSTM is applied to assess the dynamic landslide susceptibility based on multidimensional and multiscale temporal and spatial data. The aim of this research is the assessment of landslide susceptibility based on machine-learning algorithms for the China-Nepal Highway in the Hindu Kush-Himalayan region, taking into consideration the various instability factors and their evolution features.

2. Methodology

2.1. Study Area

The China-Nepal Highway, marked as an orange bold line in Figure 1, is located in the central part of the Hindu Kush Himalayan region (HKH). It runs east to west over 943 km from Lhasa, the capital of Tibet, China, to Kathmandu, capital of the Federal Republic of Nepal. The highway stretches through four large mountains, namely the Tolsan (elevation 4950) and Gatzola Mountain. (elevation 5220), Tonglashan (elevation 5324), Yaxunxiong (elevation 5627), and has an average altitude of more than 4000 m. Due to the fact that the entire area is located in the slope layers and plateau terrain of the Himalayas, the terrain, geology, hydrology, and climate along the highway are extremely complex. Surrounded by high mountains, deep valleys, steep terrain, severe mountain fragmentation, strong new structure movements, frequent earthquakes, and concentrated precipitation (annual rainfall of up to 2500 mm), the highway is heavily affected by natural hazards such as landslides, fragmentation, landslides, and mudslides.

(a)   (b)

Figure 1. (**a**) map of the China-Nepal Highway; (**b**) Location map of the study area.

The study area is located in the Nyalam of Shigatse area, where geological disasters occur most frequently. This part stretches 133 km from Mengla in the north to Friendship Bridge bordering Nepal in the south, comprising longitudes 85°57′55″–86°10′7″ and latitudes 27°58′20″–28°48′30″. The topology of the area undulates dramatically, with elevations ranging from 1770 m to 5123 m.

2.2. Instability Factors

The first problem to be addressed is the detection of instability factors which cause mountain hazards of different types and degrees. With the development of space techniques and information technologies, a great variety of temporal and spatial data become available, such as geological data, geographic information, high-resolution remote sensing images, hydrological data, and so on. These instability factors can fall into three categories: disaster-causing factors, disaster-pregnant environment factors, and hazard-bearing body factors. A disaster-pregnant environment is characterized by topography, lithology, and the formation of strata, as well as land use. Disaster-causing factors include the precipitation and dynamic change of glacial lakes. The vulnerability degree of hazard-bearing bodies and the dangerous degree of the above two factors together decides the severity of mountain hazard. Since the instability factors are numerous, and most of them have obvious fuzziness and uncertainty, it is difficult to extract key factors that can provide accurate and real-time hazard susceptibility assessment from multi-source data.

Figure 2 illustrates the landslide susceptibility assessment framework based on multi-source data integration and deep learning algorithms. Data sources related to mountain hazards include digital elevation model (DEM), high-resolution remote sensing images (HR-RS), 1: 50,000 geologic maps

(GM), and meteorological data (MD). Different features can be extracted from the aforementioned raw data, as follows.

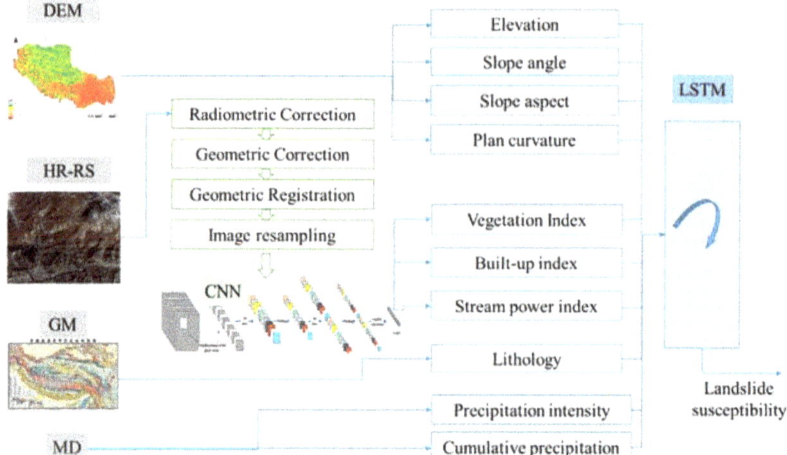

Figure 2. Framework of integrated deep learning-based landslide susceptibility assessment.

2.2.1. Features Based on DEM

Slope angle: Slope degree is one of the most frequently-used factors in assessing landslide susceptibility [13–18]. It has a great influence on slope stability and is directly related to the different types of mountain hazards (Figure 3a).

Slope aspect: It is defined as the direction of terrain surface, such as north, northeast and so on. Since hillsides orientated differently receive direct solar radiation and rainfall in different amounts, which lead to different slope topography, humidity and plant cover, the slope aspect is also accepted as a conditioning factor (Figure 3b).

Elevation: Previous records of the China-Nepal Highway hazards indicate that landslides in that area generally occur at a middle elevation (Figure 3c). This is due to the fact that a mountain at high altitudes usually has thin soil cover and a stable rocky structure, while area at low altitudes has gentle slopes, neither of which is susceptible to landslides [13,14].

Plan curvature: Curvature is defined as the change rate of slope angle with surface plane. The direction of drainage line is influenced by plan curvature types, and the river erosion is a key factor that affects the slope stability (Figure 3d).

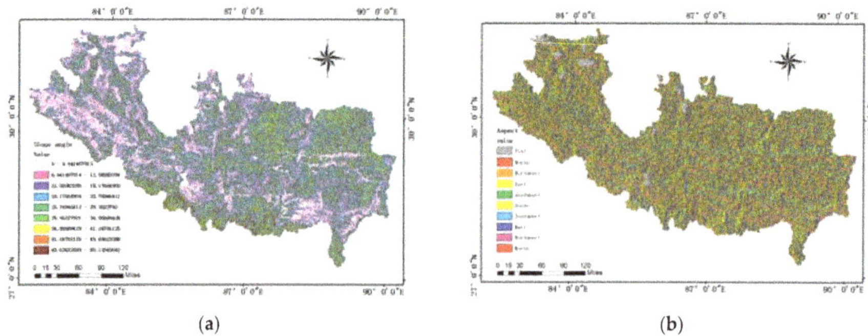

(a)

(b)

Figure 3. *Cont.*

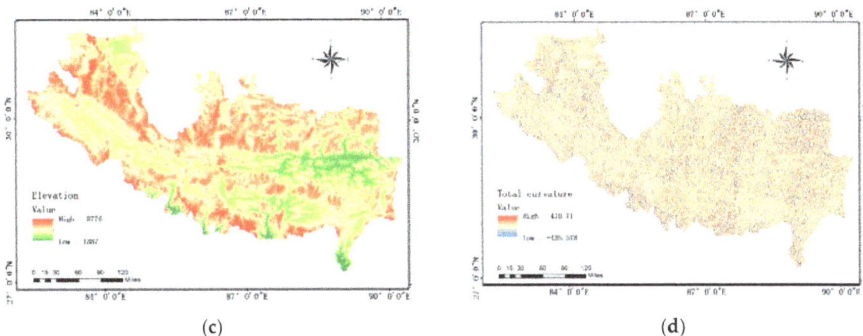

(c)   (d)

Figure 3. Spatial factors in China-Nepal highway: (**a**) slope angle; (**b**) Slope aspect; (**c**) Elevation; (**d**) Plan curvature.

2.2.2. Features Extracted from HR-RS

Remote sensing images are used to extract land cover and utilization information through object-based classification methods. A series of preprocessing work is essential for image classification, including the radiation correction, the geometric correction, the landform correction and the noise reduction. The purpose of radiation correction is to eliminate the difference of spectral reflectivity and spectral radiance between the sensor data and the real images. Geometric correction is the calibration of geometric distortions such as offset, stretching, squeezing, and distortion of the image due to factors such as the rotation or the curvature of the earth, and the temporal and spatial changes of the remote sensing platform. Then different types of land covers are classified, including water, built-up, vegetation, high-way, rock and so on. Cover area of water, vegetation and rock belong to instability factors, while built-up and high-way effect the dangerous degree of landslides. We can obtain four indicators from the classification results: vegetation index, built-up index, road index and stream power index.

2.2.3. Features Based on GM

The development of geological hazards is influenced by strata's lithology, geological structure and rock-texture. Places with strong structural deformation are easy to form folds and faults, as well as large-scale rock body rupture, which often become the solid source of landslides.

Lithology: The relationship between lithology and solid source is reflected in the weather resistance and anti-erosion ability. Generally, soft layer has low strength and weak resistance to weathering and provides more incompact solid matters. The complex geological structure and the massive loose solid materials intensifies the landslide disaster's occurring. Geology formations in the study area mainly include limestone, dolomite, sandstone and shale.

2.2.4. Features Based on MD

Water is not only an important component of landslides, but also a triggering condition and transport medium. Rainfall is an important predisposing factor in triggering landslides because it reduces soil suction and increases the pore-water pressure in soils [31–33]. Experiments have shown that the landslide occurrence is related both to the intensity and duration of a rainfall event. Thus, two indexes are used to quantify the precipitation characteristics: cumulative precipitation index (CPI) and precipitation intensity index (PII). CPI is calculated with the linear combination of antecedent precipitation in a period, while PII represents the hourly rainfalls which contributes to the landslide-triggering rainfall threshold.

$$P_{a0} = KP_1 + K^2 P_2 + \ldots + K^n P_n \tag{1}$$

P_{a0} was used to define the CPI, where P_i is the daily rainfall for the *i*-th day before day 0, *n* is the total number of days considered in the model (*n* = 10 in this work), *K* is the constant decay factor representing the outflow of the regolith (0 < *K* < 1). Figure 4 shows the changing curve of PII and CPI at an observation point during one year from 2016/01–2016/12

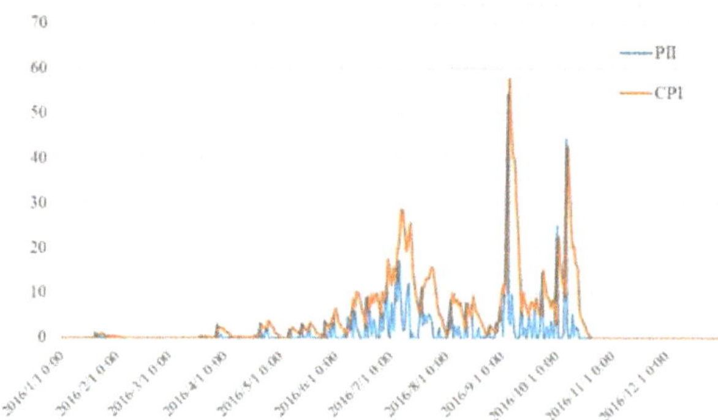

Figure 4. PII and CPI curve of an observation point from 2016/01–2016/12.

2.3. LSTM

Both SVM and NN belong to the static model, which neglect the dynamic evolution characteristics of mountains and landslide displacement and limit the improvement of prediction accuracy. Unlike the traditional neural network such as BPNN and ANN, RNN adopts recursive connection to construct its internal nodes, so that the state of the previous moment can influence the latter moment, thus realizing the state feedback of the network. However, when the information or time interval between the nodes becomes very long, "It is difficult for RNN to capture long-term time associations, which is called the "vanishing gradient problem". To solve this problem, LSTM is then proposed by adding a memory block in each unit of hidden layers, which comprises three types of gate functions—input gate, forget gate, and output gate. LSTM uses the memory mechanism to control the transmission of information at different times, which greatly improves the ability of RNN to process long-sequence data. The LSTM model structure diagram is shown in Figure 5.

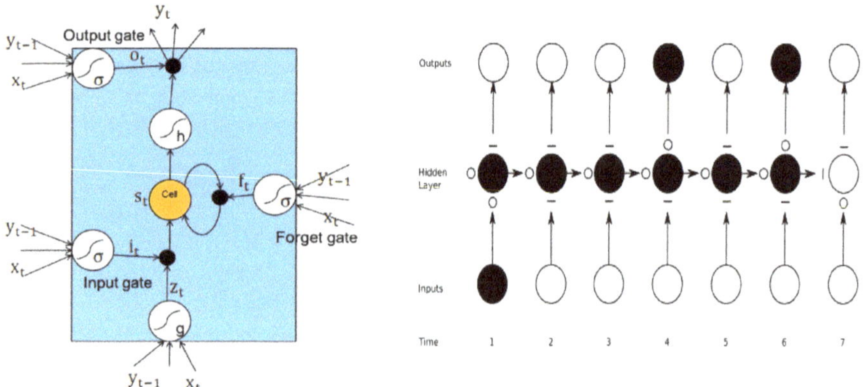

Figure 5. LSTM model structure diagram.

Input gates:

$$i_t = \sigma(W_{xi}x_t + W_{hi}h_{t-1} + W_{ci}c_{t-1} + b_i) \tag{2}$$

Forget gates:

$$f_t = \sigma(W_{xf}x_t + W_{hf}h_{t-1} + W_{cf}c_{t-1} + b_f) \tag{3}$$

Cell units:

$$c_t = f_t c_{t-1} + i_t\tanh(W_{xc}x_t + W_{hc}h_{t-1} + b_c) \tag{4}$$

Output gates:

$$o_t = \sigma(W_{xo}x_t + W_{ho}h_{t-1} + W_{co}c_t + b_o)$$
$$h_t = o_t\tanh(c_t) \tag{5}$$

where i_t, f_t, c_t, o_t represents the state vector of the input gate, forget gate, cell unit and output gate at time step t, respectively. x_t denotes the input of LSTM network at time t, W is the weight matrix between each layer, h represents the hidden state vector and b is the offset value corresponding to each gate. σ is a sigmoid activation function mapping real numbers to [0,1], while tanh is a hyperbolic tangent function mapping real numbers to [−1,1].

3. Results

3.1. Four Prediction Models

A total of 3800 data points collected from the monitoring site during the period from January 2015 to December 2016 were used in this experiment, which is shown in Figure 6. Data collected between January 2015 and June 2016 were used as a training data set, and the remaining data were used as a test data set. Data preprocessing is performed before the entire data set is split. In order to reduce the influence of the landslide evaluation factor data type, value range, and dimension inconsistency on the prediction model, the original data is normalized to [0,1] closed interval. For each of the attribute values in the evaluation factor, the attribute values are normalized, and the normalization method uniformly uses the range standardization. The sensitivity index was divided into stable, low susceptibility, moderate susceptibility, medium susceptibility and high susceptibility, and very high susceptibility. Thus, the landslide susceptibility assessment is transformed into a classification problem.

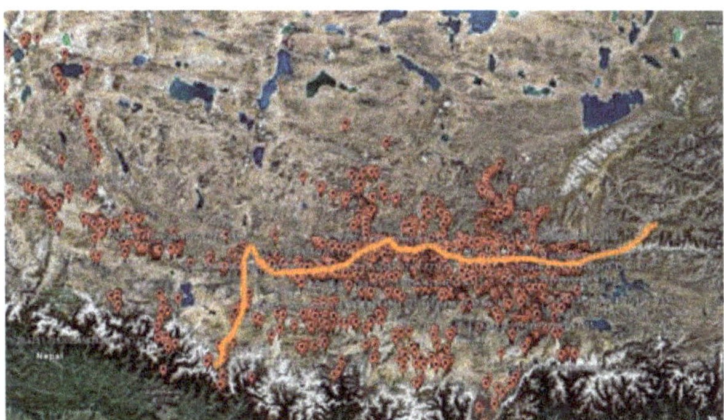

Figure 6. Sample points of test area.

Four common classification algorithms are used in the paper to compare with the LSTM model, decision tree (DT), support vector machines (SVM), and Back Propagation neural network (BPNN). The DT and BPNN prediction are performed using Matlab R2013b. LSTM and SVM are implemented in Python using the open source deep learning framework Keras package (which uses TensorFlow as a backend) and the Scikit-learn package, respectively. The parameters of these models are as follows. Table 1 shows the optimal parameters of the four models.

Table 1. Optimal parameters of different models.

Model	Parameter	Value	Description
BPNN	Number of hidden layer neurons	21	
	Activation function	Sigmoid function	
SVM	c	0.15	Penalty coefficient
	g	0.75	Parameter of RBF
	Kernel function	Radial basis functions	
DT	criterion	Gini	Criterion for feature selection
	max_depth	30	Maximum depth of the tree
	min_samples_leaf	50	Minimum sample number of the leaf node
LSTM	input sequence length	8	
	Loss function	Categorical cross-entropy	

For BPNN, the most widely-used three-layer network consists of an input layer with 10 neurons, one hidden layer with 21 neurons, and one output layer with 1 neuron; it was built as a network structure. The number of hidden layer neurons is determined according to the empirical equation $N_h = 2N_i + 1$, where N_h represents the number of hidden layer neurons and N_i is the number of input layer neurons. Since the initialization weights and thresholds of the BP network have a great influence on the training speed and effect, this paper adopts genetic algorithm to optimize these parameters.

For SVM, the kernel function is the most important factor determining the model prediction effect. The K-fold Cross Validation (K-CV) method is applied to search the optimal parameters ($K = 20$ in the paper). The original data is divided into K groups, of which each subset data is used as a test set and the remaining K-1 subset data is used as a training set. By using the K-CV method, the classification accuracies under different combination of c and g are obtained. The combination of c and g with the highest classification of accuracy is selected as the best parameter.

For DT, the purpose of parameter optimization is to prevent the structure of the tree from being too large, resulting in over-fitting problems. Info entropy and gini index are the most commonly-used impurity functions to split the nodes. max_depth and min_samples_leaf act as a constraint to determine the termination of the decision tree construction, thereby controlling the size of the tree.

For LSTM, the length of the input sequence determines the number of the historical data points in the recursive connection. By the grid search method, the input sequence length is set to 8 in this paper.

3.2. Experiment Result

As mentioned above, this paper establishes the landslide hazard prediction as a classification problem, and the sample points can be divided into six categories according to different landslide susceptibility levels, i.e., stable, low susceptibility, moderate susceptibility, medium susceptibility and high susceptibility, and very high susceptibility. In the experiment, stable is denoted as label 1, while very high susceptibility is denoted as label 6. Through expert experience and manual judgment, the number of sample points of each susceptibility level is shown in Table 2, where 1612 of 3800 points are in a stable condition, 934 of 3800 points are in a low susceptibility condition, 549 of 3800 points are in a moderate susceptibility condition, 259 of 3800 points are in a medium susceptibility condition, 234 of 3800 points are in a high susceptibility condition, and 212 of 3800 points are in a very high susceptibility condition.

Table 2 and Figure 7 illustrate the prediction results of different classification models. In Table 2, take the first row as an example; it shows that by applying BPNN model, 1015 points are correctly classified into label 1 (stable), which means the accuracy is 62.97%. For the sample points in label 1, BPNN, SVM, DT, LSTM models achieved accuracies of 62.97%, 76.36%, 64.21%, and 82.20%, respectively. Figure 7 shows the confusion matrixes of the four models. It is a visual display tool for evaluating the quality of a classification model, wherein each column of the matrix represents the sample label predicted by the model, while each row of the matrix represents the true label of the sample.

Table 2. Classification results of different models.

	Models	Label 1	Label 2	Label 3	Label 4	Label 5	Label 6
Label 1	BPNN	1015	434	115	31	13	4
(1612)	SVM	1231	351	2	10	10	8
	DT	1035	285	142	84	41	25
	LSTM	1325	251	36	0	0	0
Label 2	BPNN	154	597	90	56	30	7
(934)	SVM	204	691	16	5	13	5
	DT	179	526	94	94	25	16
	LSTM	56	801	40	19	13	5
Label 3	BPNN	67	96	312	51	19	4
(549)	SVM	12	133	384	15	5	0
	DT	32	40	322	82	50	23
	LSTM	12	83	423	23	8	0
Label 4	BPNN	3	20	49	170	15	2
(259)	SVM	0	5	54	179	21	0
	DT	16	11	31	159	23	19
	LSTM	0	4	34	198	23	0
Label 5	BPNN	0	19	21	35	131	28
(234)	SVM	0	3	38	53	140	0
	DT	10	19	20	29	132	24
	LSTM	0	1	18	23	172	20
Label 6	BPNN	1	7	19	20	33	132
(212)	SVM	0	0	0	4	64	144
	DT	19	18	17	16	20	122
	LSTM	0	0	0	14	32	166

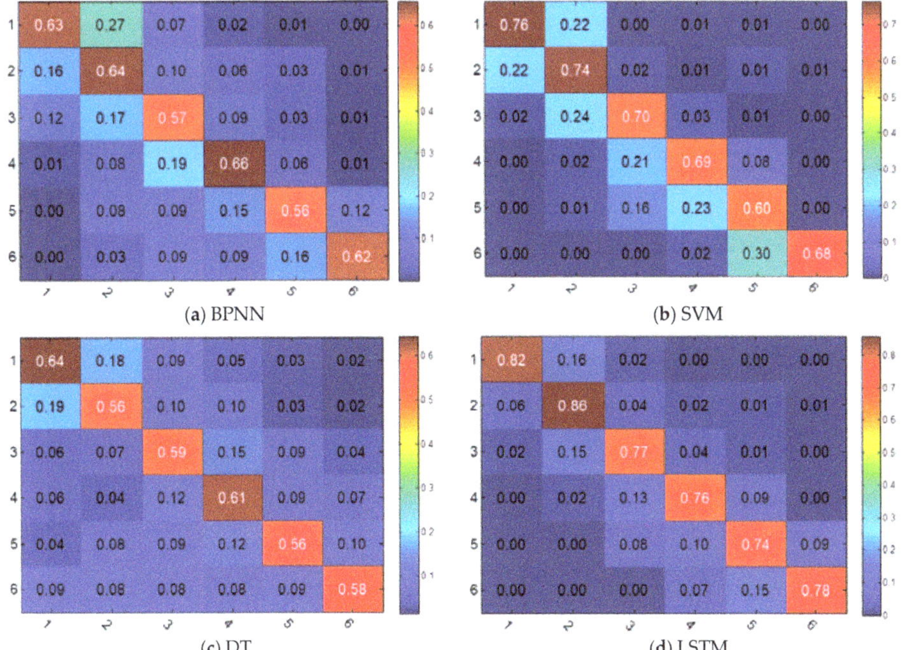

Figure 7. Multi-class confusion matrix of the four models for landslide hazard prediction.

In actual situations, prediction results of the landslide susceptibility level within a certain margin of error are acceptable. For example, if the actual area is in a stable condition and by prediction models it is classified as being in a low susceptibility category, then the prediction results can be considered as acceptable. Thus, in the paper, the prediction error level (PEL) is defined as an indicator to measure the prediction effect of different models.

$$\text{PEL}_k = \frac{\sum_{i=1}^{6} \sum_{J=\max(i-k,0)}^{J=\min(i+k,6)} \hat{N}_j}{\sum_{i=1}^{6} N_i} \tag{6}$$

where PEL_k represents the kth prediction error level, $\hat{N}_j$ is the points number in label J of the prediction results, N_i is the points number in label i of the actual sample.

In Table 3, 0-level represents the prediction accuracy of different labels, while 1-level means the prediction error is only one interval, for example, the actual condition is low susceptibility while the predicted condition is stable or moderate susceptibility. In practice, the prediction results with 0-level or 1-level error are acceptable and can be used to make preventative and control measures. In Figure 8, we can see that almost 90% of the prediction errors of LSTM are 0-level or 1-level.

Table 3. PEL results of different models.

	0-Level (Excellent)	1-Level (Good)	2-Level (Moderate)	3-Level (Poor)	4-Level (Bad)	5-Level (Very Bad)
BPNN	62.03%	25.92%	8.42%	2.97%	0.53%	0.13%
SVM	72.87%	23.97%	1.87%	0.69%	0.40%	0.21%
DT	60.42%	21.24%	10.10%	4.85%	2.24%	1.16%
LSTM	81.18%	15.40%	2.92%	0.37%	0.13%	0

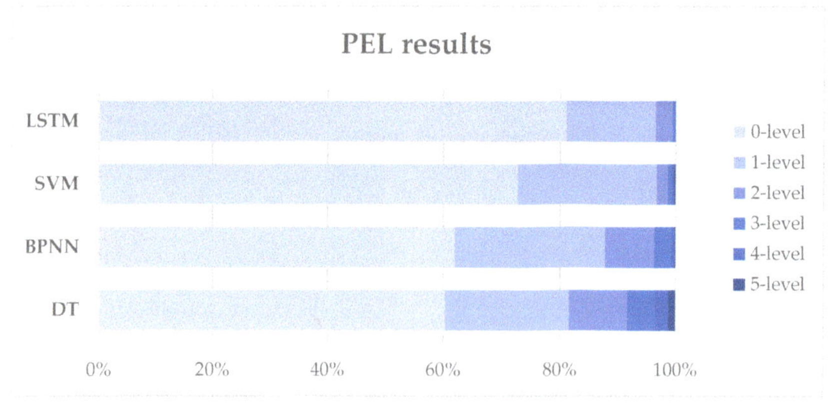

Figure 8. PEL results of the four models.

4. Discussion

The overall prediction accuracies of BPNN, SVM, DT, and LSTM are 62.0%, 72.9%, 60.4% and 81.2%, respectively. As the performance of a data-driven model is greatly affected by the sample size, there are differences in prediction accuracies among different labels. From Table 2, we can also conclude that high susceptibility is the most difficult condition to predict, since it only has an accuracy of 73.5% by LSTM, while the stable condition has an accuracy of 82.2%. In general, LSTM and SVM outperform BPNN and DT in each category in terms of stability of accuracy across different folds of the tested dataset. This is due to the fact that SVM is a structural learning method, which makes it advantageous in solving high dimensional models of small-sample sets. Meanwhile, the historical information from the previous steps contained in the hidden layer of LSTM makes it the most accurate among the four models.

The confusion matrixes in Figure 6 show that there is a certain classification error between label 1 and label 2 for BPNN, SVM, and DT, which means that it is hard for them to distinguish the low susceptibility from the stable condition. Although SVM has a relatively good accuracy, it does not perform well in classifying the neighboring two categories. From this perspective, LSTM is better than SVM since the boundary between the diagonal section and other section in the confusion matrix is obvious. From Figure 8 we can see that the prediction error at level 2 or below of all these four models accounts for more than 90%, which means the four models can predict the landslide susceptibility well within an acceptable error range. The LSTM model has the lowest probability of large prediction errors (3-level or above), while the DT model has the highest probability, which is 0.5% and 8.24%, respectively. We can also conclude that the performance of SVM model is very close to the performance of LSTM, when considering the probability of small prediction errors (1-level or below), which are 96.84% and 96.58%, respectively.

According to the prediction results, the very high susceptibility dataset has either of the following characteristics: (1) elevation higher than 4000 m, lithology with shales, slope angle from $40°$ to $55°$, and vegetation index lower than 10 (2) elevation from 2000 m to 2800 m, slope angle from $20°$ to $35°$, plan curvature higher than 200 and CPI higher than 30. This result is in accordance with the actual situation.

5. Conclusions

The China-Nepal Highway is an important part of the Belt and Road development strategy. Due to the harsh natural environment along the road, the frequency and intensity of local mountain disasters are increasing, and the casualties and economic losses are increasing accordingly. Therefore, this paper takes the China-Nepal Highway as the research object and conducts risk assessments for mountain disasters. With the development of information and sensing technology in recent years, more and more sensor data and remote sensing data are collected, and a great variety of temporal and spatial data has become available, such as geological data, geographic information, high-resolution remote sensing images, hydrological data, and so on. The influence of various factors on risk has the characteristic of ambiguity, and hierarchies exist between the various degrees of influence. Classical mathematical models are ill-suited to express these complex relationships. At the same time, previous studies only used the static data and characteristics of the study area to characterize the intensity of landslides and debris flow disasters, and these factors have dynamic evolution characteristics.

To solve this problem, a novel and dynamic model that can remember historical data using so-called "memory blocks" is proposed to solve the problem of the hysteresis effects of triggering factors and landslide susceptibility. The other three classic classification models, BPNN, SVM, and DT, are also applied for comparisons with the LSTM model in landslide susceptibility assessments. The results of this study showed that the SVM model (72.87%) had better accuracy than the BPNN (62.03%) and DT model (60.42%). The LSTM model (81.18%) outperformed SVM in prediction accuracy, and they have the similar performance when considering about the probability of small prediction errors (1-level or below).

Author Contributions: Methodology, G.P. and L.X.; software, G.P.; writing—original draft preparation, G.P. and L.X.; writing—review and editing, G.P.; funding acquisition, Y.Z.

Funding: This research was funded by the National Science Foundation of China, grant number 41661144039, and the Fundamental Research Funds for the Central Universities grant number FRF-TP-18-035A1.

Conflicts of Interest: The authors declare no conflict of interest.

References

1. Liu, P.; Li, Z.; Hoey, T.; Kincal, C.; Zhang, J.; Zeng, Q.; Muller, J.-P. Using advanced InSAR time series techniques to monitor landslide movements in Badong of the Three Gorges region, China. *Int. J. Appl. Earth Obs.* **2013**, *21*, 253–264. [CrossRef]
2. Vahidnia, M.H.; Alesheikh, A.A.; Alimohammadi, A.; Hosseinali, F. A GIS-based neuro-fuzzy procedure for integrating knowledge and data in landslide susceptibility mapping. *Comput. Geosci.* **2010**, *36*, 1101–1114. [CrossRef]
3. Nichol, J.E.; Shaker, A.; Wong, M.-S. Application of high-resolution stereo satellite images to detailed landslide hazard assessment. *Geomorphology* **2006**, *76*, 68–75. [CrossRef]
4. Marjanović, M.; Kovačević, M.; Bajat, B.; Voženílek, V. Landslide susceptibility assessment using SVM machine learning algorithm. *Eng. Geol.* **2011**, *123*, 225–234. [CrossRef]
5. Pu, F.; Ma, J.; Zeng, D.; Xu, X.; Chen, N. Early Warning of Abrupt Displacement Change at the Yemaomian Landslide of the Three Gorge Region, China. *Nat. Hazard. Rev.* **2015**, *16*, 04015004. [CrossRef]
6. Chen, W.; Peng, J.; Hong, H.; Shahabi, H.; Pradhan, B.; Liu, J.; Zhu, A.X.; Pei, X.; Duan, Z. Landslide susceptibility modelling using GIS-based machine learning techniques for Chongren County, Jiangxi Province, China. *Sci. Total Environ.* **2018**, *626*, 1121–1135. [CrossRef] [PubMed]
7. Chen, Z.; Zhang, Y.; Ouyang, C.; Zhang, F.; Ma, J. Automated Landslides Detection for Mountain Cities Using Multi-Temporal Remote Sensing Imagery. *Sensors* **2018**, *18*. [CrossRef]
8. Skilodimou, H.; Bathrellos, G.; Koskeridou, E.; Soukis, K.; Rozos, D. Physical and Anthropogenic Factors Related to Landslide Activity in the Northern Peloponnese, Greece. *Land* **2018**, *7*, 85. [CrossRef]
9. Bathrellos, G.D.; Skilodimou, H.D.; Chousianitis, K.; Youssef, A.M.; Pradhan, B. Suitability estimation for urban development using multi-hazard assessment map. *Sci. Total Environ.* **2017**, *575*, 119–134. [CrossRef]
10. Nandi, A.; Shakoor, A. A GIS-based landslide susceptibility evaluation using bivariate and multivariate statistical analyses. *Eng. Geol.* **2010**, *110*, 11–20. [CrossRef]
11. Bai, S.-B.; Wang, J.; Lü, G.-N.; Zhou, P.-G.; Hou, S.-S.; Xu, S.-N. GIS-based logistic regression for landslide susceptibility mapping of the Zhongxian segment in the Three Gorges area, China. *Geomorphology* **2010**, *115*, 23–31. [CrossRef]
12. Yilmaz, I. Comparison of landslide susceptibility mapping methodologies for Koyulhisar, Turkey: Conditional probability, logistic regression, artificial neural networks, and support vector machine. *Environ. Earth Sci.* **2009**, *61*, 821–836. [CrossRef]
13. Liu, M.; He, Y.; Wang, J.; Lee, H.P.; Liang, Y. Hybrid intelligent algorithm and its application in geological hazard risk assessment. *Neurocomputing* **2015**, *149*, 847–853. [CrossRef]
14. Xu, C.; Dai, F.; Xu, X.; Lee, Y.H. GIS-based support vector machine modeling of earthquake-triggered landslide susceptibility in the Jianjiang River watershed, China. *Geomorphology* **2012**, *145–146*, 70–80. [CrossRef]
15. Pradhan, B. A comparative study on the predictive ability of the decision tree, support vector machine and neuro-fuzzy models in landslide susceptibility mapping using GIS. *Comput. Geosci.* **2013**, *51*, 350–365. [CrossRef]
16. Pham, B.T.; Tien Bui, D.; Prakash, I.; Dholakia, M.B. Hybrid integration of Multilayer Perceptron Neural Networks and machine learning ensembles for landslide susceptibility assessment at Himalayan area (India) using GIS. *Catena* **2017**, *149*, 52–63. [CrossRef]
17. Kalantar, B.; Pradhan, B.; Naghibi, S.A.; Motevalli, A.; Mansor, S. Assessment of the effects of training data selection on the landslide susceptibility mapping: A comparison between support vector machine (SVM), logistic regression (LR) and artificial neural networks (ANN). *Geomat. Nat. Hazards Risk* **2017**, *9*, 49–69. [CrossRef]
18. Lee, S.; Hong, S.-M.; Jung, H.-S. A Support Vector Machine for Landslide Susceptibility Mapping in Gangwon Province, Korea. *Sustainability* **2017**, *9*, 48. [CrossRef]
19. Pourghasemi, H.; Gayen, A.; Park, S.; Lee, C.-W.; Lee, S. Assessment of Landslide-Prone Areas and Their Zonation Using Logistic Regression, LogitBoost, and NaïveBayes Machine-Learning Algorithms. *Sustainability* **2018**, *10*, 3697. [CrossRef]
20. Pourghasemi, H.R.; Kerle, N. Random forests and evidential belief function-based landslide susceptibility assessment in Western Mazandaran Province, Iran. *Environ. Earth Sci.* **2016**, *75*. [CrossRef]

21. Pradhan, B.; Lee, S. Landslide susceptibility assessment and factor effect analysis: Backpropagation artificial neural networks and their comparison with frequency ratio and bivariate logistic regression modelling. *Environ. Model. Softw.* **2010**, *25*, 747–759. [CrossRef]
22. Tsangaratos, P.; Ilia, I. Comparison of a logistic regression and Naïve Bayes classifier in landslide susceptibility assessments: The influence of models complexity and training dataset size. *Catena* **2016**, *145*, 164–179. [CrossRef]
23. Sezer, E.A.; Pradhan, B.; Gokceoglu, C. Manifestation of an adaptive neuro-fuzzy model on landslide susceptibility mapping: Klang valley, Malaysia. *Expert Syst. Appl.* **2011**, *38*, 8208–8219. [CrossRef]
24. Aghdam, I.N.; Pradhan, B.; Panahi, M. Landslide susceptibility assessment using a novel hybrid model of statistical bivariate methods (FR and WOE) and adaptive neuro-fuzzy inference system (ANFIS) at southern Zagros Mountains in Iran. *Environ. Earth Sci.* **2017**, *76*. [CrossRef]
25. Ghorbanzadeh, O.; Rostamzadeh, H.; Blaschke, T.; Gholaminia, K.; Aryal, J. A new GIS-based data mining technique using an adaptive neuro-fuzzy inference system (ANFIS) and k-fold cross-validation approach for land subsidence susceptibility mapping. *Nat. Hazards* **2018**, *94*, 497–517. [CrossRef]
26. Razavi Termeh, S.V.; Kornejady, A.; Pourghasemi, H.R.; Keesstra, S. Flood susceptibility mapping using novel ensembles of adaptive neuro fuzzy inference system and metaheuristic algorithms. *Sci. Total Environ.* **2018**, *615*, 438–451. [CrossRef] [PubMed]
27. Chalkias, C.; Ferentinou, M.; Polykretis, C. GIS Supported Landslide Susceptibility Modeling at Regional Scale: An Expert-Based Fuzzy Weighting Method. *ISPRS Int. J. Geo-Inf.* **2014**, *3*, 523–539. [CrossRef]
28. Ma, X.; Tao, Z.; Wang, Y.; Yu, H.; Wang, Y. Long short-term memory neural network for traffic speed prediction using remote microwave sensor data. *Transport. Res. Part C Emerg. Technol.* **2015**, *54*, 187–197. [CrossRef]
29. Yu, J.J.Q.; Hill, D.J.; Lam, A.Y.S.; Gu, J.; Li, V.O.K. Intelligent Time-Adaptive Transient Stability Assessment System. *IEEE Trans. Power Syst.* **2018**, *33*, 1049–1058. [CrossRef]
30. Mezaal, M.R.; Pradhan, B.; Sameen, M.I.; Mohd Shafri, H.Z.; Yusoff, Z.M. Optimized Neural Architecture for Automatic Landslide Detection from High-Resolution Airborne Laser Scanning Data. *Appl. Sci.* **2017**, *7*, 730. [CrossRef]
31. Ma, T.; Li, C.; Lu, Z.; Wang, B. An effective antecedent precipitation model derived from the power-law relationship between landslide occurrence and rainfall level. *Geomorphology* **2014**, *216*, 187–192. [CrossRef]
32. Brunetti, M.T.; Peruccacci, S.; Rossi, M.; Luciani, S.; Valigi, D.; Guzzetti, F. Rainfall thresholds for the possible occurrence of landslides in Italy. *Nat. Hazards Earth Syst. Sci.* **2010**, *10*, 447–458. [CrossRef]
33. Li, C.; Ma, T.; Zhu, X.; Li, W. The power–law relationship between landslide occurrence and rainfall level. *Geomorphology* **2011**, *130*, 221–229. [CrossRef]

MDPI

St. Alban-Anlage 66

4052 Basel

Switzerland

Tel. +41 61 683 77 34

Fax +41 61 302 89 18

www.mdpi.com

Sensors Editorial Office

E-mail: sensors@mdpi.com

www.mdpi.com/journal/sensors